DNA Sequencing III: Dealing with Difficult Templates

Edited By
Jan Kieleczawa, PhD
Wyeth Research, Cambridge, Massachusetts

JONES AND BARTLETT PUBLISHERS
Sudbury, Massachusetts
BOSTON   TORONTO   LONDON   SINGAPORE

World Headquarters
Jones and Bartlett Publishers
40 Tall Pine Drive
Sudbury, MA
01776
978-443-5000
info@jbpub.com
www.jbpub.com

Jones and Bartlett Publishers
Canada
6339 Ormindale Way
Mississauga, Ontario L5V 1J2
CANADA

Jones and Bartlett
Publishers International
Barb House, Barb Mews
London W6 7PA
UK

Jones and Bartlett's books and products are available through most bookstores and online booksellers. To contact Jones and Bartlett Publishers directly, call 1-800-832-0034, fax 978-443-8000, or visit our website www.jbpub.com

Production Credits
Chief Executive Officer: Clayton Jones
Chief Operating Officer: Don W. Jones, Jr.
President, Higher Education Professional Publishing: Robert W. Holland, Jr.
V.P., Design and Production: Anne Spencer
V.P., Sales and Marketing: William Kane
V.P., Manufacturing and Inventory Control: Therese Connell
Executive Editor: Cathleen Sether
Acquisitions Editor: Shoshanna Goldberg
Managing Editor: Dean W. DeChambeau
Associate Editor: Molly Steinbach
Senior Production Editor: Louis C. Bruno, Jr.
Production Assistant: Leah Corrigan
Marketing Manager: Andrea DeFronzo
Text Design: Louis C. Bruno, Jr.
Cover Design: Anne Spencer
Composition: SNP Best-Set Typesetter, Ltd., Hong Kong
Printing and Binding: Malloy
Cover Printing: Malloy

Library of Congress Cataloging-in-Publication Data
DNA sequencing III : dealing with difficult templates / Jan Kieleczawa.—1st ed.
   p. ; cm.
Includes bibliographical references and index.
ISBN 978-0-7637-4297-3 (alk. paper)
   1. DNA—Analysis.   2. Nucleotide sequence—Methodology.   I. Kieleczawa, Jan.
   II. Title: DNA sequencing three.   III. Title: DNA sequencing 3.
[DNLM: 1. Sequence Analysis, DNA–methods.   2. Base Sequence.   3. Templates,
   Genetic. QU 450 D62971 2008]
QP624.D17492 2008
572.8′6—dc22   2007042261
6048

Printed in the United States of America
12 11 10 09 08   10 9 8 7 6 5 4 3 2 1

Brief Contents

Contents

Contributors

Mostafa Ait-Zahra
Biological Technologies Department, Wyeth Research, Cambridge, Massachusetts
Chapter 5

Vinay Dhodda
Lucigen Corporation, Middleton, Wisconsin
Chapter 7

Carl W. Fuller
GE Healthcare, Piscataway, New Jersey
Chapter 4

Ronald Godiska
Lucigen Corporation, Middleton, Wisconsin
Chapter 7

Rebecca Hochstein
Lucigen Corporation, Middleton, Wisconsin
Chapter 7

Attila Karsi
College of Veterinary Medicine, Mississippi State University, Mississippi State, Mississippi
Chapter 7

Jan Kieleczawa
Biological Technologies Department, Wyeth Research, Cambridge, Massachusetts
Chapters 1, 2, 5, 8

Aaron Kitzmiller
Analysis Development and Application Services, Wyeth Research, Cambridge, Massachusetts
Chapter 8

Donald Koffman
Biological Technologies Department, Wyeth Research, Cambridge, Massachusetts
Chapter 8

Bharath Lakshmanan
Biological Technologies Department, Wyeth Research, Cambridge, Massachusetts
Chapter 8

Tony Li
Biological Technologies Department, Wyeth Research, Cambridge, Massachusetts
Chapter 5

David Mead
President, Lucigen Corporation, Middleton, Wisconsin
Chapter 7

Richard McCombie
Department of Psychiatry, The Mount Sinai School of Medicine of New York University, and Psychiatry Research, Bronx Veterans Administration Medical Center, Bronx
Foreword

Nikolai Ravin
Center "Bioengineering," Russian Academy of Science, Moscow, Russia
Chapter 7

Masanori Suzuki
Team Leader, Transcriptional Regulatory Network Exploration Team, RIKEN Genomic Sciences Center, Yokohama, Japan
Chapter 6

Chengcang Wu
Lucigen Corporation, Middleton, Wisconsin
Chapter 7

Paul Wu
Inflammation Department, Wyeth Research, Cambridge, Massachusetts
Chapter 5

Haiguang Xiao
GE Healthcare, Piscataway, New Jersey
Chapter 4

Alicia Yang
Applied Biosystems, Inc., Foster City, California
Chapter 3

Preface

This is the third volume in the DNA Sequencing series published by Jones and Bartlett. *DNA Sequencing: Optimizing the Process and Analysis*, 2005 and *DNA Sequencing II: Optimizing Preparation and Cleanup*, 2006 have a number of chapters about the optimization of almost every step in the DNA sequencing process. The "classical" Sanger approach to DNA sequencing is a very mature technology with enormous success (complete sequencing of hundreds of microbial species and numerous higher organisms). Over the past few years, however, new technologies seem to be taking the spotlight in the DNA sequencing field with already acknowledged advances coming from 454 Life Sciences/Roche (GS 20 and GS FLX systems) and the already-available-for-commercial-use instruments from Illumina (1G Analyzer), Applied Biosystems (SOLiD), Helicos, and a number of other manufacturers. All of these systems, and others soon to come, hold enormous promise to reduce sequencing costs and to increase the number of organisms to be sequenced or resequenced. The eventual goal is to deliver a human-size genome at a cost of $1000 or less. In this volume, however, we focus entirely on the existing Sanger sequencing chemistry.

For thousands of existing DNA sequencing facilities, the novel technologies are still quite in the distant future, and one may reasonably expect that many of them will not be able or will not have the need to harness their potential power. The current capillary technology is ideal for sequencing (or resequencing templates after some modifications) even a few thousand individual clones. Applied Biosystems, and possibly other vendors, is committed to supporting their current line of capillary sequencers. However, it is quite unlikely that there will be any further technological developments to improve capillary sequencing instruments. The investment in many research centers will be in improving chemistries, tweaking various protocols to increase read lengths, and dealing more effectively with unusual templates.

This volume is entirely devoted to the Sanger sequencing chemistry, specifically how to deal with various difficult DNA templates. In our experience, the difficult templates constitute only about 5% to 8% of all templates submitted for sequencing. If the DNA sequencing staff lacks the expertise and tools to deal with such situations, however, the delivery of good quality data can be prevented or much longer turnaround times for data delivery can result, because of the many trials necessary to produce any reasonable data. For many years, for example, the most common approach to sequencing a GC-rich template was to include 5% DMSO, 1 M betaine, or to substitute pure BigDye™ with a 4 : 1 v/v mixture of BigDye:dGTP in the sequencing reaction (see Chapters 1 and 2). Sometimes these modifications resulted in improved data quality, but recently more systematic studies of the effects of various treatments for different categories of difficult templates were undertaken and published. Although we are still far away from establishing firm rules (for example, if the template is over 80% GC-rich, then treatment A will always work), the incorporation of controlled heat-denaturation steps together with various additives considerably improved chances of getting good quality data for many kinds of difficult templates.

Chapter 1 deals with heat denaturation of plasmids and PCR fragments and its effect on the quality of sequencing data. The new aspect in this work (compared to formerly published material) is the addition of the data for heat denaturation of plasmids with difficult regions and for PCR fragments of various lengths. The incorporation of heat-denaturation steps almost always improves read length and the quality of data for all kinds of plasmids. The overall read length for heat-denatured PCR fragments is only slightly improved (compared to samples without heat denaturation), but the quality of data is significantly better, especially for PCR fragments with many mixed bases. Above all, the unified treatment of mixed templates simplifies handling of sequencing reactions.

Chapter 2 presents extensive data on the effects of heat denaturation in combination with other additives on the quality of almost 60 different difficult templates. These (and the data presented in Chapters 3 and 4) are likely the most extensive collection of templates and treatments ever assembled in a single publication.

For many years, Applied Biosystems scientists presented at conferences many suggestions for sequencing various difficult templates, but this is the first time that a more detailed paper has been released for general use. The cornerstone of Chapter 3 is the introduction of a few dye-terminator formulations that can be used to sequence through specific categories of difficult templates. Hopefully, these dye formulations will soon be available for general use. It is quite possible that the combination of heat denaturation and any given formulation will further improve the sequence quality of many difficult templates.

Chapter 4 describes application of the Sequence Resolver Kit (from GE Healthcare) to sequence many types of difficult templates. This approach seems most useful for GC-rich templates and those with hairpins and with various repeats, but it is not effective, for instance, in templates with long poly A/T stretches. It is a very valuable tool, however, in any DNA-sequencing facility that routinely deals with various difficult templates.

Quite often, published sequences of vectors, or those posted on a vendor's Web site, are derived from *in silico* cloning and stitching pieces together without real sequence confirmation. But if experiments do not work as expected, the scientists may want to know the true sequence of a vector to explain reasons for deviations or failures. Chapter 5 describes detailed sequence analysis of some pDEST vectors and the undocumented presence of a very strong hairpin. In fact, the differences between real and posted sequences are quite dramatic for pDEST series (from Invitrogen). We strongly advise that the sequence of any vector that is routinely used in experiments be verified to make sure that all-important features are present and there are no major deviations that could lead to incorrect conclusions. In addition, a number of other hairpin structures have been sequenced using many different chemistries, and the correlation between melting temperature (T_m) and the "easiness" of the clean sequence trace has been established.

Chapter 6 describes a novel approach to sequencing difficult regions, based on transcriptional sequencing. Despite its limitations (currently validated only on ABI 377), it can be a valuable tool for those who still possess this kind of instrumentation. On the other hand, we hope that such an approach could be validated for other kinds of sequencing analyzers, therefore, expanding its utility.

DNA sequencing of a template is often one of the steps in the long process of getting and characterizing a desired clone. There are considerable challenges in getting clones with some difficult features (e.g., with very AT-rich, GC-rich, or highly repetitive regions). Chapter 7 describes an elegant system for cloning and characterizing templates that are otherwise unclonable in many standard systems.

The development and application of various biochemical methods represents only one aspect of sequencing many difficult templates and it is mostly applied after the first, standard approach fails or results in low-quality data. The other, complementary, approach is to apply bioinformatics tools to a reference sequence (if known) to discover whether such a sequence contains potentially difficult-to-sequence regions. Almost all primer design commercial packages (Primer Designer, Oligos, DNAStar, etc.) contain algorithms that will point out a limited number of potentially difficult-to-design primers or difficult-to-sequence stretches, for example, homopolymers, inverted repeats, or hairpins. However, the lack of integration with other laboratory information management systems (LIMS) (such as those needed to create DNA sequencing requests and sample sheets) and limited options make their use quite cumbersome and ineffective. Chapter 8 describes DNA LIMS (used to submit/process sequencing requests and to store finished results) that incorporate GC and repeats modules. GC module calculates GC% separately in both forward and reverse directions and has associated chemistries based on the level of the GC content built in the LIMS. The repeat module (named Examine Repeats) scans the molecule of choice for seven different, potentially difficult-to-sequence, regions (homopolymers, direct and inverted repeats, di-and tri-nucleotides, long stretches of di-nucleotides that do not form repeats, and regions that could form compressions). Though we have developed chemistries appropriate for each type of difficult region, they are not currently automatically linked in our LIMS system (meaning that if a template contains, for instance, an inverted repeat, then LIMS will apply/suggest chemistry A).

We hope that this volume will be an invaluable source of articles and methods for all those who routinely or occasionally need to fully sequence difficult DNA templates. We also believe that, to further our understanding of the correlation between successful sequencing outcomes and a specific chemistry, one should closely analyze, using sophisticated bioinformatics tools, the sequence that directly precedes and follows the difficult region. This will require much closer collaboration between experienced bench scientists and bioinformaticians.

Acknowledgments

I was truly lucky to find a few contributors whose outstanding knowledge of the topic was essential to the entire endeavor. Thank you so much for your contributions, help, and understanding during the long, exhausting, yet rewarding process of creating this book. Your contributions added many other angles to the project and helped to create a manual that hopefully will be useful to many who often struggle with those pesky and unruly DNA templates.

I would like to thank the entire staff of the DNA Sequencing Group at Biological Technologies of Wyeth Research who often fed me their troublesome templates; without those templates many chapters would be less informative. The truly outstanding staff at Jones and Bartlett Publishers, particularly Lou Bruno, Cathleen Sether, Shoshanna Grossman, Molly Steinbach, Shellie Newell, Jan Cocker, and Sherri Dietrich, deserve all the credit for the efficient and speedy production of this book. It's a pleasure to collaborate with all of you and I hope to continue this relationship for years to come. Finally, I wish to thank my family, Carla, Michael, Alex, and Kasia for all their support and understanding.

Jan Kieleczawa
Groton, Massachusetts
March 2008

Foreword

I first began using automated sequencers in 1988. They were a far cry from today's instruments. There were neither BigDye™ terminators nor cycle sequencing reactions. As automated sequencing developed hand in hand with the human genome project, the protocols available became more robust and provided better results. But the focus of the genome became the rapid and cheap sequencing of the easy parts of the genome, and less focus was given to sequencing the difficult parts. For the genome project, this made a lot of sense. For many other applications, however, it made much less sense. Because the great push of the genome project drove the technology, however, sequencing difficult regions never got much attention.

Now we are at another technological inflection point. So-called "next-generation sequencing technologies" are rapidly replacing ABI3730 capillary sequencers for bulk sequencing of the majority of the targets of sequencing, but again, not the difficult regions. In the very near future the capillary sequencers will be used more and more for confirmation of resequencing data from next-generation sequencers and filling in problem areas. This may very well be some of the difficult regions in the latter case. So it seems likely that a major part of the future use of capillary sequencers will be in sequencing difficult templates. This book is interesting for two reasons as a result of this trend: It is first a compilation of what we have learned about sequencing difficult regions over the years, and, more importantly, it is a starting point for the future use of capillary-based sequencers as they shift from a high-throughput role in the lab to a more focused and, in some ways, more demanding role. This book provides an outstanding starting point for their new use.

Richard McCombie
Cold Spring Harbor Laboratory
March 2008

1 Controlled Heat-Denaturation of DNA Plasmids and PCR Fragments

Jan Kieleczawa
Wyeth Research, Cambridge, MA

Unless the DNA template to be sequenced is single-stranded (ss DNA), the two strands in plasmid DNA (ds DNA) and in polymerase chain reaction (PCR) templates must be separated for the priming event and extension to occur. Prior to the introduction of thermostable DNA polymerases in the sequencing process (9, 10), the strand separation was accomplished by heat-denaturation (22° to 100°C) of plasmids in the presence or absence of 0.1 to 0.3 N NaOH, followed by neutralization, alcohol precipitation, and subsequent resuspension in the desired solution (3–5, 13, 16). Though effective, these steps are cumbersome, time-consuming, and in this author's opinion, far from optimal. In fact, most denaturation conditions described in publications (3–5, 13, 16, and the references therein) were replicated and data evaluated using agarose gel electrophoresis and DNA sequencing. In all cases, the transition from ds to ss form is only partial, if any, and additional nonsequenceable bands are formed (as shown in Figure 1-1), thus effectively reducing the amount of a template available during the cycling step. Table 1-1 shows the sequence data for four different NaOH-induced plasmid denaturation protocols and the denaturation conditions recommended in this work.

The results presented in this section could be valuable to those who still use manual radioactive plasmid DNA sequencing with non-thermostable polymerases where the initial strand separation is an essential step. Based on the data from the renaturation experiments (the denatured template stays in ss form for quite a long time), it also would be relatively easy to incorporate such a controlled heat-denaturation step into any process flow in high-throughput DNA sequencing centers that

DNA III: Dealing with Difficult Templates
Edited by Jan Kieleczawa
©2008 Jones and Bartlett Publishers

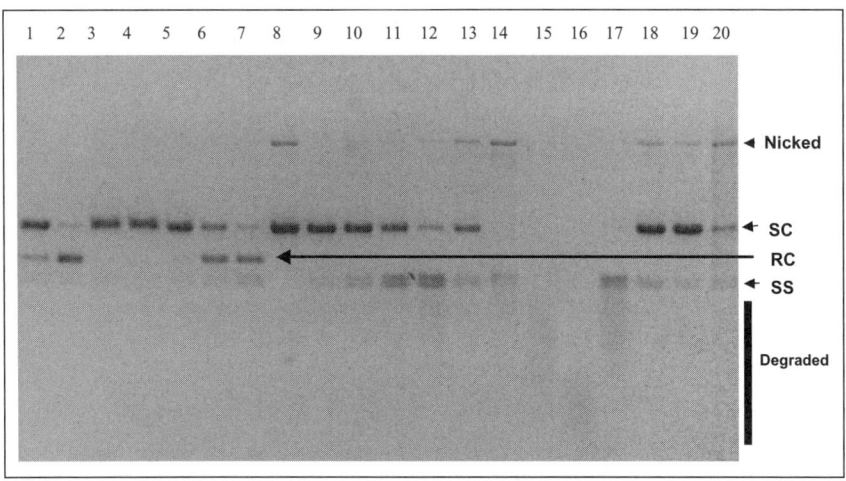

Figure 1-1. Denaturation of plasmid DNA under various conditions. Aliquots of 200 ng of pGem3zf were denatured using several published protocols (lanes 1–7) and using heat-denaturation at elevated temperatures as described in reference 4 (lanes 9–16). Lane 1 = denaturation at 0.1 N NaOH at room temperature for 5 minutes (as in reference 4). Lane 2 = as lane #1 but denaturation at 0.2 N NaOH (as in reference 1). Lane 3 = denaturation in the presence of 0.2 M EDTA at 37°C (as in reference 13). Lane 4 = as lane #3 but denaturation at 85°C (as in reference 13). Lane 5 = denaturation in the presence of 0.1 N NaOH for 30 minutes at 37°C. Lane 6 = denaturation in the presence of 0.1 N NaOH for 5 minutes at 85°C. Lane 7 = denaturation in the presence of 0.2 N NaOH for 3 minutes at 100°C (as in reference 16). Lane 8 = control DNA, no heat denaturation. Lanes 9–12 = heat denaturation at 98°C in 10 mM Tris/0.01 mM EDTA for 1, 2.5, 5 and 10 minutes, respectively. Lanes 13–16 = heat denaturation at 98°C in water for 1, 2.5, 5, and 10 minutes, respectively. Lane 17 = pGem3zf in 10 mM Tris/0.01 mM EDTA after 25 sequencing cycles as described under DNA sequencing section. Lane 18 = pGem3zf in 2 mM MgCl$_2$ after 25 sequencing cycles as described in Experimental Design. Lane 19 = pGem3zf in the presence of BigDye terminator V3.0 (final MgCl$_2$ = 2 mM) after 25 sequencing cycles as described under DNA sequencing section. Lane 20 = pGem3zf subjected first to 5 minutes of heat denaturation in 10 mM Tris/0.01 mM EDTA followed by 25 sequencing cycles in the presence of BigDye terminator V3.0 (final MgCl$_2$ = 2 mM) as described under DNA sequencing section.

rely heavily on automation. As shown in Chapter 2, adding a controlled heat denaturation step to the sequencing protocol is one of the most effective ways to sequence through many different categories of difficult DNA templates. In addition, it is possible to use this protocol in general PCR technology to improve the quality of the product and to further reduce the amount of the initial DNA template. Any other DNA technology that relies on effective strand separation could benefit from this protocol.

Table 1-1. Comparison of read lengths and signal strength for varying DNA concentrations and different denaturation protocols.

Conditions ng DNA		This work/H_2O[1]		This work/TE_{sl}[2]		Ref. 3	Ref. 11[3]	Ref. 11[4]	Ref. 14
		−HD	+HD	−HD	+HD				
25	Q ≥ 20	489 ± 87	744 ± 53	488 ± 84	724 ± 71	268 ± 62	284 ± 124	223 ± 92	0
	SS	13 ± 4	32 ± 6	12 ± 2	27 ± 6	9 ± 2	11 ± 4	9 ± 2	6 ± 1
50	Q > 20	585 ± 177	764 ± 44	625 ± 106	784 ± 59	577 ± 44	409 ± 59	436 ± 92	339 ± 90
	SS	16 ± 5	47 ± 9	17 ± 3	46 ± 20	13 ± 3	11 ± 2	9 ± 2	8 ± 1
200	Q > 20	837 ± 25	843 ± 22	724 ± 71	866 ± 29	792 ± 39	663 ± 113	621 ± 76	712 ± 41
	SS	60 ± 10	119 ± 14	42 ± 7	123 ± 20	65 ± 2	43 ± 10	30 ± 8	26 ± 6

Varying amounts of pGem3zf plasmid DNA (4 samples for each condition) were pretreated differently before subjecting them to cycle sequencing. Processed sequencing reactions were run on ABI3100 equipped with an 80-cm capillary array. The data were evaluated in terms of Q ≥ 20 read length (RL) and signal strength (SS in fluorescent units).

[1] DNA was resuspended in water and sequenced using standard protocol (−HD = no heat denaturation), or samples were heat denatured (+HD) for 2.5 minutes at 98°C before cycle sequencing.

Ref. 3: DNA was denatured in 0.1N NaOH at room temperature for 5 minutes and neutralized with 0.1N HCl.

[2] DNA was resuspended in 10 mM Tris/0.01 mM EDTA and sequenced using standard protocol (−HD = no heat denaturation), or samples were heat denatured (+HD) for 7.5 minutes at 98°C before cycle sequencing.

[3] DNA was denatured at room temperature (22°C) for five minutes in the presence of 0.2N NaOH. Following this treatment, equal concentration of HCl was added to neutralize this solution. The mixtures were supplemented with 40mL of water, 6mL of 3 M sodium acetate pH 6.0, and 160mL 95% ethanol, and then were centrifuged for 30 minutes at 3400 × g; Supernatants were discarded, and precipitates were washed twice with 200mL of 70% ethanol. Finally, samples were dried for 15 minutes at 65°C and resuspended in 6mL water, 1mL of 5-mM M13 reverse primer, and 3 μL of twofold diluted BigDye terminator V3.0 and electrophoresed as described above.

[4] As above, but the denaturation was at 85°C.

Ref. 14: Denaturation was in 0.1N NaOH at 100°C followed by ethanol precipitation and resuspension.

Compared to the prior work (6), this chapter significantly expands most sections and adds new sections on kinetics of renaturation of plasmids and PCR fragments after the heat denaturation.

Experimental Design

All heat-denaturing experiments were carried out in a PTC-225 thermocycler (BioRad, Hercules, CA) in 200 µL PCR tubes covered with caps. Unless otherwise mentioned, the denaturation step was performed on about 100 to 200 ng of each studied plasmid DNA or PCR fragment in a final volume of 10 µL in 10 mM Tris/Cl pH 8, 0.01 mM EDTA (TE$_{sl}$), or in water. During the preparation, sample tubes were stored on ice. All of the tubes were then placed in a preheated thermocycler block set at 98°C, unless otherwise stated. At specified time intervals, tubes were quickly withdrawn from the cycler and placed on ice until the end of the series. Samples were briefly centrifuged and 1.1 µL of 10 × DNA loading buffer was added and the entire sample loaded onto a 1% agarose gel (1 × TAE buffer/0.5 µg/mL ethydium bromide, agarose gel size 14 × 11 cm) as described in (11). Samples were electrophoresed for approximately two hours at 100 to 150 V. Incubation in water for 30 minutes (in a cold room) removed excess ethydium bromide. The intensities of the fluorescently stained bands were measured and quantified using EDAS 290 Kodak 1D Image Analysis Software (Eastman Kodak Company, Rochester, NY). The time at which 75% of the ds DNA is converted to ss DNA was determined for each plasmid from their spectral characteristics. Initially, similar experiments were carried out in 0.1, 0.2, and 0.3 N NaOH. However, the need for careful neutralization, possible precipitation steps, quite ineffective conversion to ss form as well as the formation of additional DNA bands (Figure 1-1) rendered such an approach impractical, and it was therefore abandoned.

DNA Sequencing

Unless otherwise indicated, the DNA sequencing was carried out as follows. An aliquot (0.2 µg or as indicated in a specific experiment) of plasmid DNA (or 25 ng of PCR fragment or as indicated in specific experiments) was combined with 1 µL of 5 µM primer and 3 µL of twofold diluted ABI PRISM BigDye Terminator Cycle Sequencing Ready Reaction Kit mix (version 3.0 or 3.1). The volumes were adjusted to 10 µL with 10 mM Tris-HCl, 0.01 mM EDTA (pH 8.0) or water, and amplification reactions were performed on a PTC-225 cycler (BioRad) for 40 cycles (96°C for 10 seconds, 50°C for 5 seconds, and 60°C for 2 minutes). Twenty microliters of water were added to the reactions and the excess dye was removed by gel filtration on a Performa® DTR V3 96-well filter plate (purchased from EdgeBiosystems, Gaithersburg, MD). If other cleanup protocol was used, it is so indicated in a specific legend. The samples were heat-denatured for two minutes at 90° to 95°C and electrophoresed on ABI3100 or ABI3730 Genetic Analyzers (Applied Biosystems, Foster City, CA) under conditions recommended by the manufacturer.

Temperature-Dependence of Plasmid Denaturation

In TE_{sl} hardly any transition from the ds to ss form occurs at temperatures below 80°C. Figure 1-2 shows an example of two plasmids denatured at various temperatures. It is apparent that any appreciable ds to ss conversion takes place only at temperatures above 90°C. For practical reasons, 98°C was selected for further experiments; however, slightly lower or higher temperatures can be used with proper time adjustments that can be estimated from Figure 1-2.

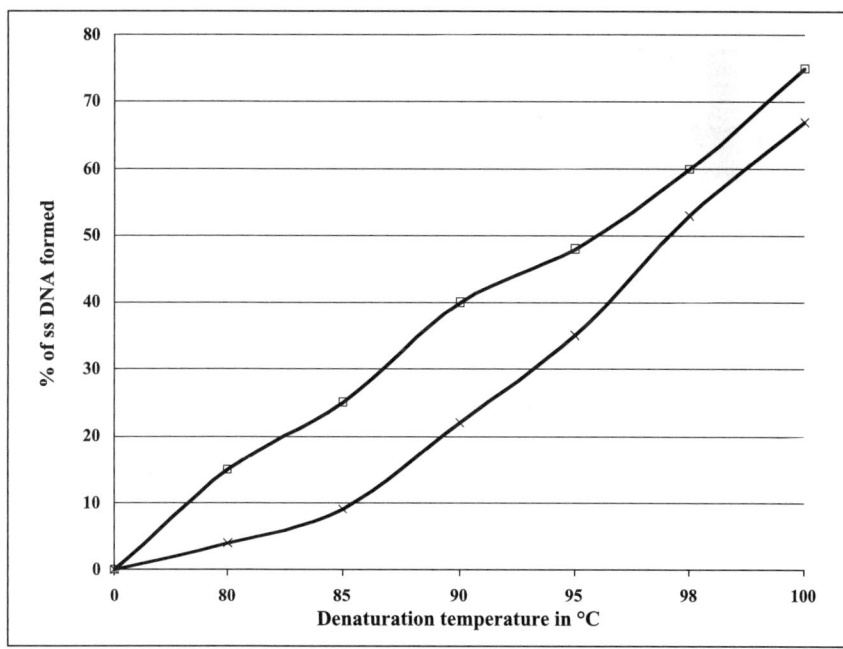

Figure 1-2. **Temperature-dependence of plasmid denaturation.** Two hundred nanograms of plasmids pGem3zf (3.2 kbp; ×) and pf2549 (16 kbp; □) were subjected to heat-denaturation for five minutes at the indicated temperatures. All samples were then processed as described in the text.

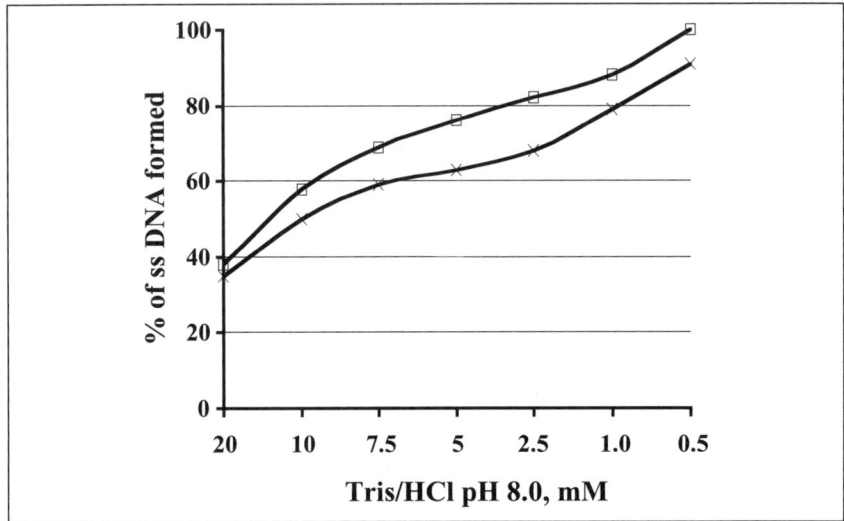

Figure 1-3. Buffer strength-dependence of plasmid denaturation. Two hundred nanograms of plasmids pGem3zf (3.2 kbp; ×) and pf2549 (16 kbp; □) was subjected to heat-denaturation for five minutes at 98°C in indicated buffers. All samples were then processed as described in the text.

Buffer-Strength Dependence of Plasmid Denaturation

The amount of salts present during heat-denaturation affects the time needed for optimal transition from ds to ss DNA. Figure 1-3 shows a denaturation experiment for two different plasmids under varying salt conditions (0.5 to 20 mM Tris/Cl buffer pH 8.0). It is worth noting that in a set period of time, almost twice as much DNA is converted from ds to ss in 5 mM Tris/Cl compared to 20 mM Tris/Cl. As a rule of thumb, the denaturation time in water or in the presence of salts below 1 mM is about half that required in 10 mM Tris buffer. Because we recommend storing plasmid DNA in 10 mM Tris/0.01 mM EDTA, we selected these conditions as the default for all subsequent experiments. Studies (8, 12, 14, 15) also compare spectral characteristics of DNAs stored in water or in buffered solutions.

Time Course of Heat-Denaturation of DNA Plasmids of Varying Sizes

A series of high-quality DNA plasmids with a size range of 3 to 20 kbp was subjected to heat denaturation at 98°C as described above. Plotting the time points at which 75% of each plasmid was converted from ds to ss form vs. the log of DNA mass resulted in a linear relationship, shown

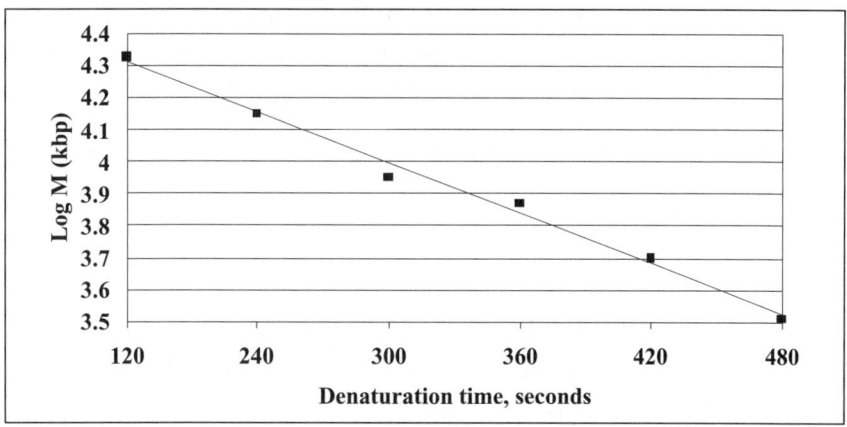

Figure 1-4. **Size-dependence of plasmid denaturation.** Two hundred nanograms of each of the following plasmids were denatured in a time-dependent manner: pGem3zf (3.2 kbp), pRJR1 (4.84 kbp), M13mp18 RF (7.25 kbp), p380 (9.0 kbp), FES7 (14.0 kbp) p2000 (20.8 kbp). Similar linear data was obtained for plasmids denatured in 0.2 N NaOH and for different levels of denaturation (not shown).

in Figure 1-4. The following experimental equation was derived from this graph:

$$DT_{75\%} \text{ (min)} = 7.5 - 1[(A - 3.2)/2.5]$$

where $DT_{75\%}$ is the denaturation time at which 75% of ds DNA is converted into ss form; A is the size (in kbp) of the plasmid to be denatured; 3.2 is the size of pGem3zf (in kbp); and 2.5 is the factor derived from Figure 1-4. It refers to the fact that for any plasmid bigger than pGem3zf, one needs to subtract 1 minute per multiple of 2.5 kbp from 7.5 minutes to achieve a similar level of denaturation. The time at which 75% of ds DNA is converted into the ss form was selected for safety, as continuing the denaturation for longer times will lead to the degradation of the DNA.

The EDTA at 1 mM substantially inhibits the transition from ds to ss form; above 5 mM EDTA, no transition at all is observed. Similar inhibitory effects are seen with NaCl above 50 mM and $MgCl_2$ above 1 mM (not shown).

For plasmids of very similar sizes, it is possible to precisely calculate the time needed for the desired level of denaturation. However, in a typical DNA sequencing core laboratory, one very rarely deals with such an idealized situation. Plasmids to be sequenced most often range in size from 4 to 10 kbp. It would be impractical to individually denature plasmids in each size range; therefore, we recommend using an average denaturation time of three to five minutes. If plasmids are delivered in water, one needs to halve the denaturation time compared to the time for DNA stored in TE_{sl}. However, the denaturation of plasmids in water leads to the formation of additional "non-sequenceable" DNA bands (see Figure 1-1). Although effective and used in many laboratories, it is not highly recommended based on the data presented in this work. In addition, long-term storage of plasmids in water leads to depurination (2) and lowers the DNA sequence quality (8).

Denaturation of Plasmids Containing a Difficult Region

The denaturation data presented above relate to typical plasmids. When plasmids containing high GC-rich, CTT-rich or A/T homopolymers regions were subjected to similar denaturation experiments, the time needed for effective conversion of ds to ss form was on the order of 20 to 30 minutes. Figure 1-5 shows the denaturation profiles of few plasmids

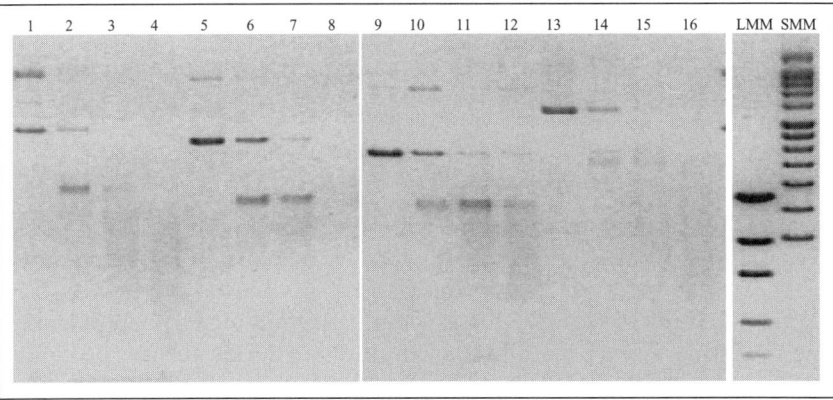

Figure 1-5. Time course of heat denaturation for plasmids containing difficult regions. Approximately 200 ng of four different plasmids were heat denatured for various periods of time and the entire aliquot was processed as described under experimental design section. Lanes 1–4 correspond to a plasmid with a region containing about 75% GC-rich region. Lanes 5–8 correspond to a plasmid with a region containing about 85% GC-rich region. Lanes 9–12 correspond to a plasmid with a region containing about 95% GC-rich region. Lanes 13–16 correspond to a plasmid with a region containing 456 base long non-repeat di-nucleotide C/T region. Abbreviations are: LMM, linear molecular mass markers (range 10 to 2000 bases); SMM, supercoiled mass markers (range 3–16 kbp). Both mass markers were purchased from Invitrogen (Carlsbad, CA). Lanes 1, 5, 9, and 13 are controls, no heat denaturation. Lanes 2, 6, 10, and 14 are samples denatured for 10 minutes. Lanes 3, 7, 11, and 15 are samples denatured for 20 minutes. Lanes 4, 8, 12, and 16 are samples denatured for 30 minutes. All denaturations were carried out in TE_{sl}.

with difficult regions. To establish a relationship between the length of heat denaturation and level of conversion to the ss form (similar to those presented in Figure 1-4 and an equation derived for "normal" plasmids) is not feasible at this time, as one would have to have the same type of difficult region present in plasmids of various sizes.

If needed, for example, when the same difficult plasmid is used routinely, its denaturation profile should be determined experimentally under laboratory-specific conditions following the guidelines described in this chapter and in reference 6.

Amount of DNA Needed for Optimal Sequencing

To determine the amount of DNA needed for optimal read length, three plasmids in the size range 3.2 to 17.2 kbp were sequenced with DNA amount ranging from 1 to 1000 ng. Each plasmid was sequenced using both standard and modified protocols with heat denaturation times adjusted as described above. Figure 1.6 shows the results of this titration (see also Figure 1-15 for additional data with different ABI instruments and in a broader DNA amount range). Without heat denaturation step, one needs 50 to 100 ng to achieve the optimal read length. With heat denaturation, only 25 to 50 ng of DNA is sufficient regardless of the size of plasmid (at least in the tested plasmid sizes).

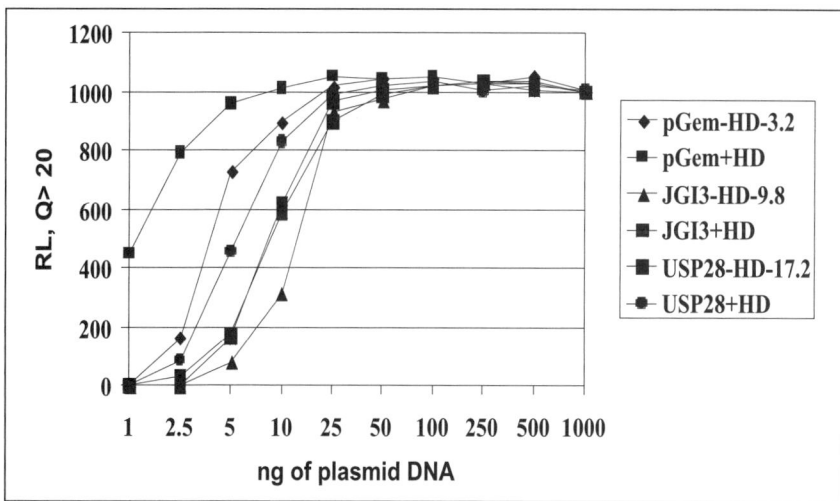

Figure 1-6. Read length versus the amount of DNA for plasmids of various sizes. Three plasmids (pGem3zf = 3.2 kbp; JG13 = 9.8 kbp; USP28 = 17.2 kbp) were sequenced using different amounts of DNA with both standard and modified protocols. The −/+ HD indicates that DNAs were sequenced without or with heat denaturation.

Note: It is worthwhile to remember that the specific numeric values for read length and signal strength depend on the type of sequencing instrument and purification used.

Renaturation Kinetics of Plasmids After Heat Denaturation

The renaturation kinetics of heat-denatured plasmids is a very slow process. Number of plasmids ("normal" and those containing high GC, CTT repeats, poly A/T homopolymers or hairpin structures) were subjected to heat denaturation (in TE_{sl} or in water) and then allowed to renature for various periods of time either in RT or on ice, and in the absence or presence of different $MgCl_2$ concentrations. Figure 1-7 shows the renaturation kinetics of pGem3zf and a pStat4 plasmid containing 19-bp hairpin structure. Figure 1-8 describes in greater detail renaturation of pGem3zf under varying $MgCl_2$ concentrations that were followed for up to four days. It is apparent that the heat-denatured samples can be left at room temperature (RT) or on ice, even for a few days, without substantial conversion to their supercoiled form. This has very practical implications, as samples can be heat denatured and left on ice or RT for extended periods of time without any measurable reversal to the pre-denatured state. However, precaution must be exercised to prevent loss of sample volume due to the evaporation, which occurs at the hourly rate of about 10% at 22.5°C or 16% at 25°C. There is no observable evaporation of samples left on ice, even for about three hours (Figure 1-9).

Denaturation of PCR Fragments of Various Sizes and Correlation with DNA Sequencing

In general, a heat-denaturation step is not necessary for PCR fragments and linear DNAs. For PCR fragments in the size range 0.6 to 3.2 kbp, 30 seconds of heat denaturation at 98°C/10 mM Tris-HCl, 0.01 mM EDTA is sufficient to separate strands to ss form. In fact, the denaturation is almost complete within 5 seconds for PCR fragments longer than 2 kbp (Figure 1-10 (a and b)). However, even 10 minutes of heat-denaturation under the same conditions has very little (if any) effect on the quality and read length of sequencing data, regardless of the length of PCR fragment (Figures 1-11 and 1-12). The initial experiments (detailed data not shown) comparing read length, signal strengths, and the overall quality for PCR fragments containing mixed bases (SNPs) indicates slightly longer reads (2%–5%) and stronger signal strength (40%–60%) are observed when a five-minute heat denaturation step is part of the sequencing protocol. But the most significant improvement is in the number of edits necessary to

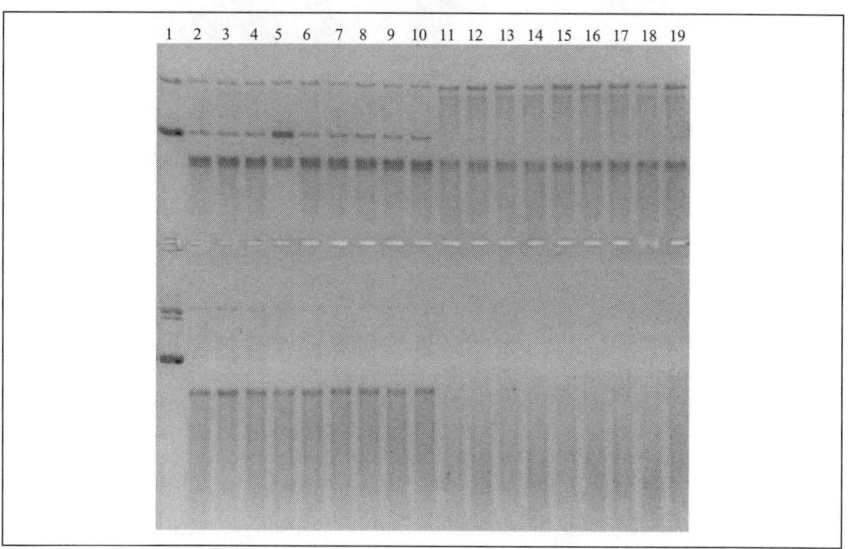

Figure 1-7. Kinetics of renaturation of pGem3zf and pStat4 plasmid containing 19-bp hairpin structure. Two hundred nanograms of DNAs were denatured for 7.5 minutes either in TE_{sl} or for 3.5 minutes in water and allowed to renature under different conditions as described below. Top part of the figure shows data for pGem3zf and the bottom part shows data for pStat4. All renaturations were carried out without magnesium ions. Lane 1 = controls, no heat denaturation. Lane 2 = 0 minutes renaturation at RT/TE_{sl}. Lane 3 = 1 hour renaturation at RT/TE_{sl}. Lane 4 = 2 hours renaturation at RT/TE_{sl}. Lane 5 = 4 hours renaturation at RT/TE_{sl}. Lane 6 = 6 hours renaturation at RT/TE_{sl}. Lane 7 = 1 hour renaturation on ice/TE_{sl}. Lane 8 = 2 hours renaturation on ice/TE_{sl}. Lane 9 = 4 hours renaturation on ice/TE_{sl}. Lane 10 = 6 hours renaturation on ice/TE_{sl}. Lane 11 = 0 minutes renaturation at RT/H_2O. Lane 12 = 1 hour renaturation at RT/H_2O. Lane 13 = 2 hours renaturation at RT/H_2O. Lane 14 = 4 hours renaturation at RT/H_2O. Lane 15 = 6 hours renaturation at RT/H_2O. Lane 16 = 1 hour renaturation on ice/H_2O. Lane 17 = 4 hours renaturation on ice/H_2O. Lane 18 = 4 hours renaturation on ice/H_2O. Lane 19 = 6 hours renaturation on ice/H_2O.

obtain clean consensus sequence in a contig: up to 36% fewer edits are needed for samples that included heat denaturation step compared to those that were sequenced using standard sequencing protocol (in one example 150 edits vs. 195 for PCR fragment of 7.75 kbp; the observed range was 1.5%–36% fewer edits). Inclusion of heat denaturation step also has a very practical implication, as there is no need to separately treat plasmid and PCR DNAs when they are handled, for example, in the same plate, or in general, in the commonly established sequencing pipeline.

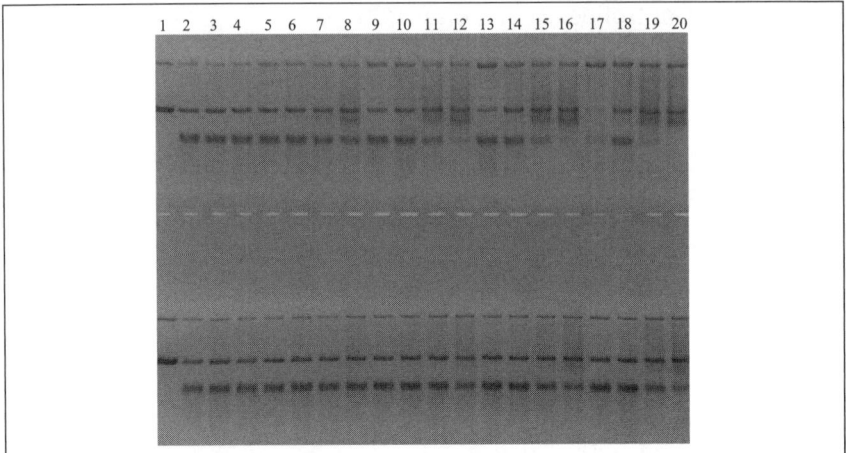

Figure 1-8. **Time-course of pGem3zf renaturation under different conditions.**
Two hundred nanograms of pGem3zf DNA were denatured for 5 minutes at
98°C/TE$_{sl}$ and allowed to renature in the absence or in the presence of various
concentrations of MgCl$_2$. Renaturation was carried out at RT (top part of the
figure) or on ice (bottom part of the figure). Lane 1 = control, no heat denaturation.
Lane 2 = 0 minutes renaturation, no MgCl$_2$. Lane 3 = 0 minutes renaturation,
5 mM MgCl$_2$. Lane 4 = 0 minutes renaturation, 10 mM MgCl$_2$. Lane 5 = 6 hours
renaturation, no MgCl$_2$. Lane 6 = 6 hours renaturation, 2 mM MgCl$_2$. Lane 7 = 6
hours renaturation, 5 mM MgCl$_2$. Lane 8 = 6 hours renaturation, 10 mM MgCl$_2$. Lane
9 = 1 day renaturation, no MgCl$_2$. Lane 10 = 1 day renaturation, 2 mM MgCl$_2$. Lane
11 = 1 day renaturation, 5 mM MgCl$_2$. Lane 12 = 1 day renaturation, 10 mM MgCl$_2$.
Lane 13 = 2 days renaturation, no MgCl$_2$. Lane 14 = 2 days renaturation, 2 mM MgCl$_2$.
Lane 15 = 2 days renaturation, 5 mM MgCl$_2$. Lane 16 = 2 days renaturation,
10 mM MgCl$_2$. Lane 17 = 4 days renaturation, no MgCl$_2$. Lane 18 = 4 days renaturation, 2 mM MgCl$_2$. Lane 19 = 4 days renaturation, 2 mM MgCl$_2$. Lane 20 = 4 days
renaturation, 10 mM MgCl$_2$.

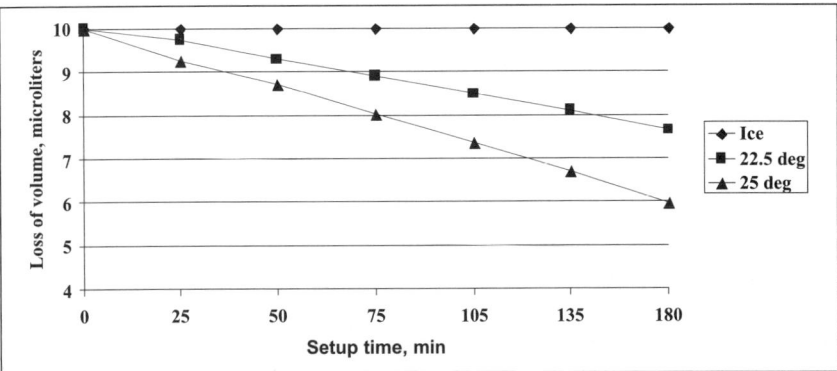

Figure 1-9. **Loss of volume under different storage conditions.** Three 10-μL
aliquots per storage conditions were pipetted into uncapped 200-μL thin wall PCR
tubes and placed on ice (♦) or in PCR blocks with temperatures set at 22.5°C (■)
and 25°C (▲). The loss of volume was measured with a carefully calibrated pipette
at indicated time intervals. The standard deviation of each measurement did not
exceed 5% to 10% of the average.

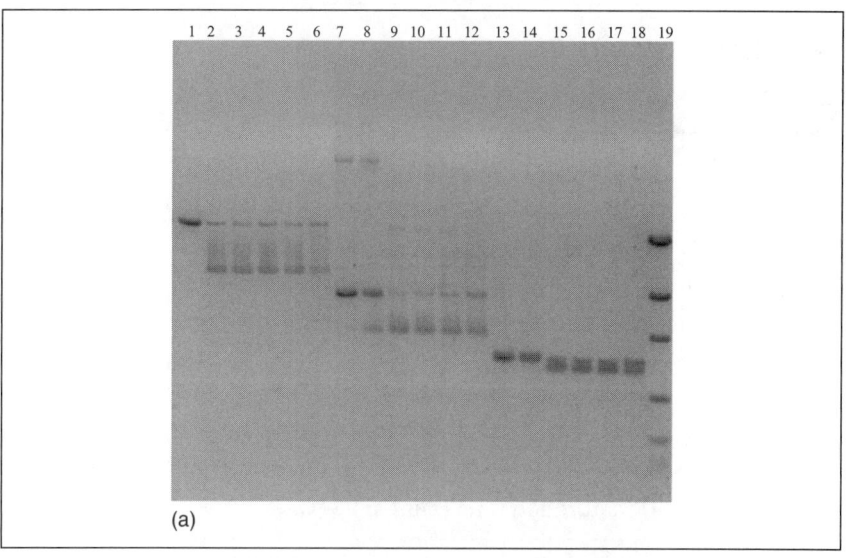

(a)

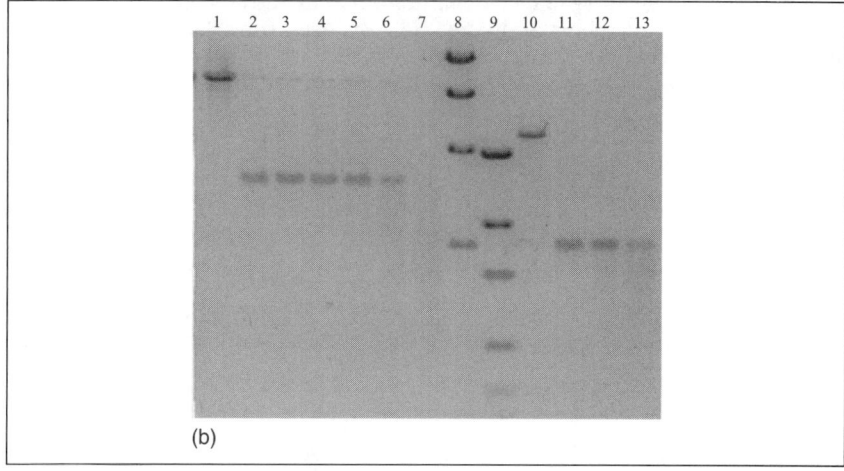

(b)

Figure 1-10. Time-course of heat denaturation of PCR fragments: denaturation profiles. (a) Three different PCR fragments were subjected to heat denaturation at 98°C in TE_{sl} and processed as described under the experimental design section in the text. Lanes 1–6 correspond to a 2.1 kbp PCR fragment. Lanes 7–12 correspond to a 1.1 kbp PCR fragment. Lanes 13–18 correspond to a 0.6 kbp PCR fragment. Lane 19 = the linear molecular mass marker as described in Figure 1.5. Lanes 1, 7, and 13 are controls (no heat denaturation). Lanes 2, 8, and 14 are samples that were heat denatured for 5 seconds. Lanes 3, 9, and 15 are samples heat denatured for 30 seconds. Lanes 4, 10, and 16 are samples heat denatured for 1 minute. Lanes 5, 11, and 17 are samples heat denatured for 5 minutes. Lanes 6, 12, and 18 are samples heat denatured for 10 minutes. (b) Two different PCR fragments, a 3.2 kbp PCR fragment and a 2.1 kbp PCR fragment, which were 84% GC-rich over the entire length were subjected to heat denaturation as described in Figure 1.8 (top). Lanes 1–7 correspond to 3.2 kbp PCR fragment. Lanes 10–13 correspond to 2.1 kbp GC-rich PCR fragment. Lane 8 = linear high molecular mass marker (purchased from Invitrogen, Carlsbad, CA, USA). Lane 9 = mass marker as described in Figure 1-5. Lanes 1 and 10 = controls (no heat denaturation). Lane 2 was sample heat denatured for 5 seconds. Lane 3 was sample heat denatured for 15 seconds. Lane 4 was sample heat denatured for 30 seconds. Lane 5 and 11 are samples that were heat denatured for 1 minute. Lanes 6 and 12 are samples heat denatured for 5 minutes. Lanes 7 and 13 are samples heat denatured for 10 minutes.

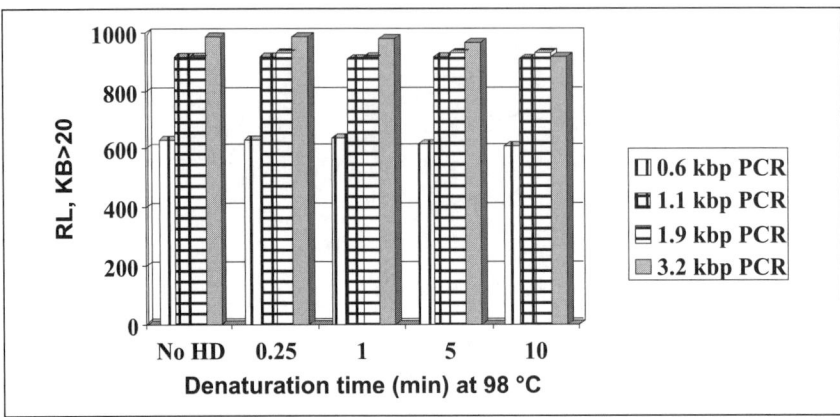

Figure 1-11. Time-course of heat denaturation of PCR fragments: effect on read length. Four PCR fragments of different lengths were subjected to heat denaturation from 15 seconds to 10 minutes to determine if there is any effect on length and quality of reads as expressed in $Q \geq 20$ values. Each data point is an average of four reads and the standard deviation is in the range of 0.2% to 4% of the average.

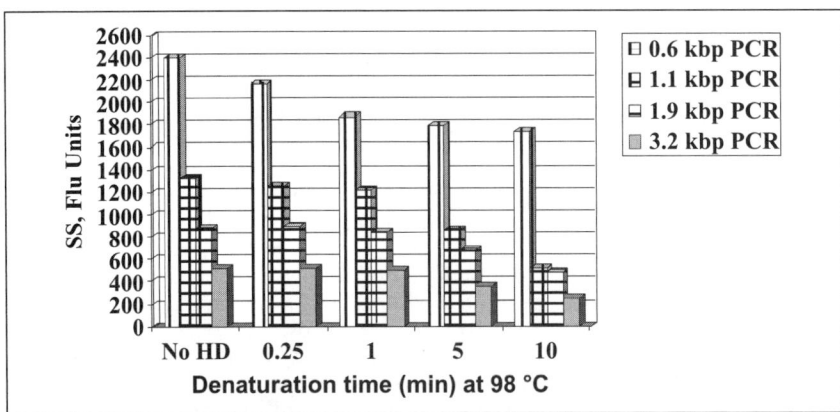

Figure 1-12. Time-course of heat denaturation of PCR fragments: effect on signal strength. Four PCR fragments of different lengths were subjected to heat denaturation from 15 seconds to 10 minutes to determine effect on signal strength. It is worth noting that signal strength is about 40 relative fluorescent units for capillaries run without any DNA (on ABI 3730). Once the signal strength exceeds 80–100 units (and is lower than 3000–5000 units), there is hardly any effect on the quality of the data and the read length. So, even as the signal strength decreases when the denaturation time increases, there is no effect on the quality and read length of sequencing data. Each data point is an average of four reads and the standard deviation is in the range of 6% to 21% of the average.

Buffer-Strength and Temperature Dependence of PCR Fragments Denaturation

As described above, the degree of plasmid denaturation depends on the strength of the buffer in which the plasmid is resuspended. In similar experiments, two PCR fragments (0.6 kbp and 3.2 kbp) were resuspended in Tris/HCl buffer, pH 8.0, at concentrations between 0.5 and 20 mM and then subjected to heat denaturation for one minute at 98°C. As shown in Figure 1-13, there is no measurable effect of buffer strength on the kinetics of conversion of ds PCR fragments to ss form, at least in the indicated buffer-strength range.

The same two PCR fragments were subjected to denaturation at different temperatures, from 60° to 98°C. From the limited data set, as expected, the denaturation of PCR fragments (carried out in TE$_{sl}$) occurs at much lower temperatures compared to those of plasmids. From the practical point of view the conversion from ds to ss form is fully

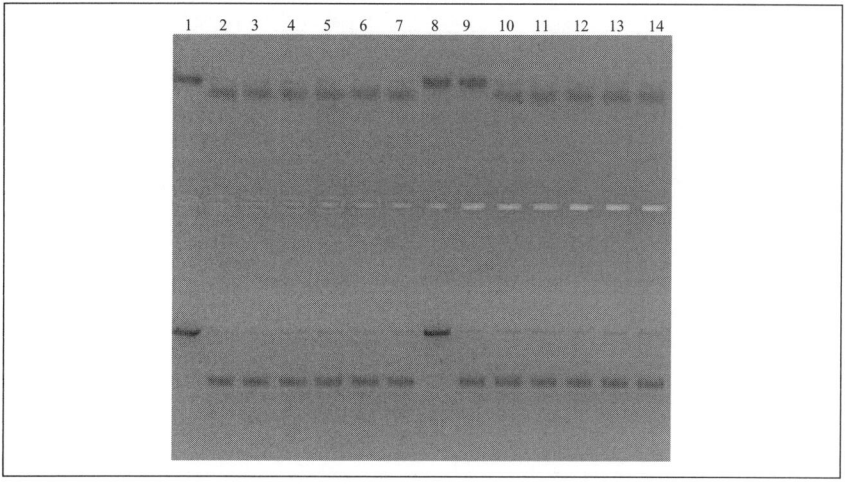

Figure 1-13. The denaturation of PCR fragments: buffer-strength and temperature dependence. In lanes 1 through 7,200 ng DNA aliquots of 0.6 kbp (top) and 3.2 kbp (bottom) PCR fragments were adjusted with Tris/HCl, pH 8.0, buffer to final concentrations between 0.5 and 20 mM and then subjected to heat denaturation for 1 minute at 98°C in 10 μL volume. All products were treated and electrophoresed under the conditions described above. Lane 1 = control (no heat denaturation), lane 2 = 0.5 mM, lane 3 = 1 mM, lane 4 = 2.5 mM, lane 5 = 5 mM, lane 6 = 10 mM, and lane 7 = 20 mM. In lanes 8 through 14, 200 ng DNA aliquots were resuspended in 10 mM Tris/HCl buffer, pH 8.0, and subjected to 1-minute heat denaturation at the temperatures ranging from 60° to 98°C. Lane 8 = 60°C, lane 9 = 70°C, lane 10 = 80°C, lane 11 = 85°C, lane 12 = 90°C, lane 13 = 95°C, and lane 14 = 98°C.

complete within 15 seconds at 98°C for PCR fragments in the size range 06 to 3.2 kbp.

Renaturation Kinetics of PCR Fragments After Heat Denaturation

Four different PCR fragments (in the size range 0.6 to 3.2 kbp) were subjected to heat denaturation in TE_{sl} for five minutes and then allowed to renature at RT or on ice for up to six hours in the absence or presence of 2 mM $MgCl_2$ (only for a 6-hour time data point). No observable reversal to the original state is visible even after six hours of renaturation when samples were left either at RT or on ice. However, in the presence of 2 mM $MgCl_2$, around 60% of the denatured 0.6 kbp PCR fragment reversed to the original form. The degree of reversal was much smaller (10%–20%) for longer PCR fragments (Figure 1-14). Again, from a practical point of view, this does not have any appreciable consequences on the DNA sequencing results.

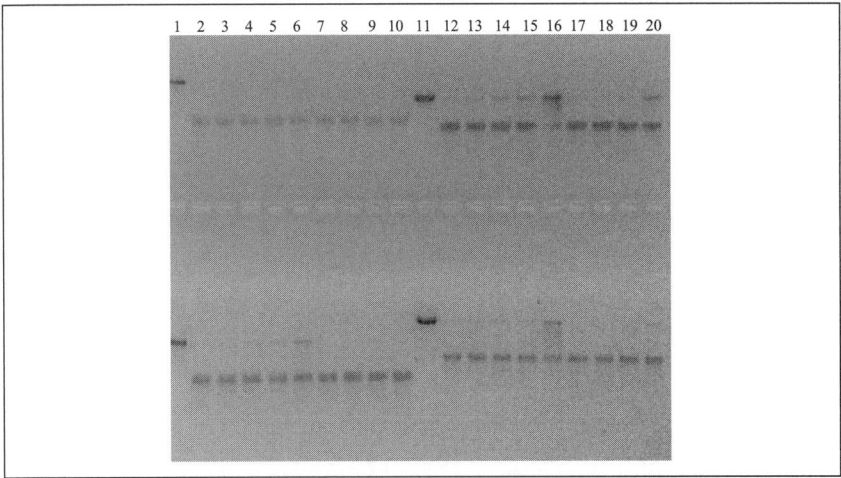

Figure 1-14. **Kinetics of renaturation of PCR fragments.** Four PCR fragments (range 0.6–3.2 kbp) were heat denatured for 5 minutes in 10 mM Tris/HCl, 0.01 mM EDTA buffer, pH 8.0, and allowed to renature either on ice or at room temperature for up to 6 hours. All renaturations, except for the 6-hour data point, were carried out without magnesium ions. Lanes 1–10 (top) = 1.1 kbp PCR fragment. Lanes 11–20 (top) = 0.6 kbp PCR fragment. Lanes 1–10 (bottom) = 1.5 kbp PCR fragment and lanes 11–20 (bottom) = 3.2 kbp PCR fragment. Lanes 1 and 11= controls (no heat denaturation). Lanes 2 and 12 = 0 minutes, lanes 3 and 13 = 1 hour, lanes 4 and 14 = 3 hours, lanes 5 and 15 = 6 hours, and lanes 6 and 16 = 6 hours in the presence of 2 mM magnesium ions. Lanes 1 to 6 and 11 to 16 were kept at room temperature and lanes 7 to 10 and 17 to 20 were kept on ice.

Benefits of Heat-Denaturation

There are a number of benefits to heat-denaturation of plasmid DNAs:

1. Less DNA is needed to obtain optimal read lengths. During the denaturation step, almost all supercoiled and nicked forms of the DNA template are converted to the sequenceable ss form. Figure 1-15 shows an example of the titration of the amount of DNA (pGem3zf) versus read length using the standard protocol and the protocol with a heat-denaturation step. DNA sequencing reactions were run on an ABI377 DNA sequencer (Applied Biosystems). For comparison, a similar titration is shown for heat-denatured samples run on an ABI3100 with a 50-cm capillary array. Note that two- to threefold less DNA is needed when a heat-denaturation step is included. For the ABI3100, a much broader usable concentration range is observed. This is especially advantageous in core laboratory settings where processed DNA samples are rarely prepared using one standard protocol and a broad range of DNA concentrations are common. Similar titration results (see Figure 1-6) were obtained on a number of plasmids in the 3 to 17 kbp size range.

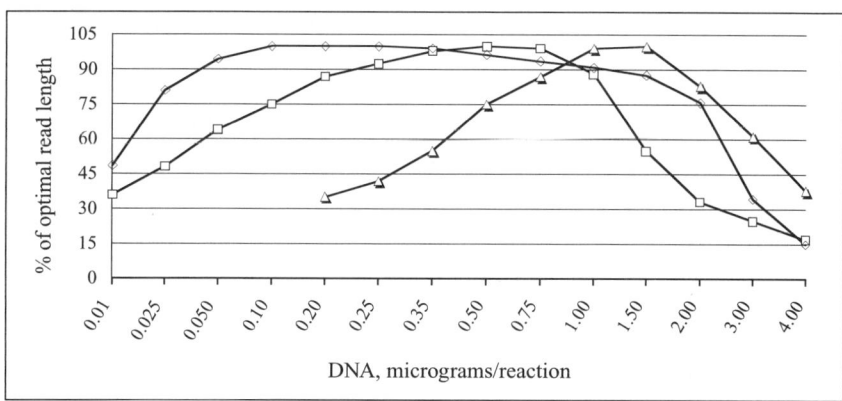

Figure 1-15. **Amount of DNA versus read length: effect of heat-denaturation.** Varying amounts of pGem3zf (run with M13 reverse primer) were sequenced using the standard (△) or modified protocol (□) and samples were run on either an ABI377 DNA sequencer or an ABI3100 (◇). For DNA run on an ABI3100, only samples prepared with the modified protocol are shown.

2. Sequencing of previously "unsequenceable" plasmids. Sometimes, inclusion of the heat-denaturation step is the only possible way to get clearly readable signal (refer to Chapter 2).
   We have encountered at least two scenarios:
   a. No chromatogram was produced using the standard protocol and clear data, though sometimes only in the range of 300 to 500 bases, when a heat-denaturation step is included. An example of such data is presented in Chapter 2, Table 2-1.
   b. Unreadable chromatograms are obtained when using the standard protocol and clear data when a heat-denaturation step is included (Figures 1-16 a–f). (See Plate 1 [Figure 1-16a] in the Color Addendum as an example.)
3. Removing most of the secondary structures in DNA. This is the most likely reason that no chromatograms were produced in case number 1 (above), as some regions in this template are over 90% GC-rich.
4. More uniform and longer read lengths. Sixteen identical samples of pGem3zf (200 ng of DNA) were subjected to sequencing using standard or modified protocols. The average signal strength (in relative fluorescent units) and the read length ($Q \geq 20$) for standard protocol were 50 ± 22 and 559 ± 230, respectively. For the modified protocol (5-min heat denaturation at 98°C), these values were 100 ± 40 and 742 ± 13, respectively. These values were obtained when the purification of sequencing products was done using in-house prepared G-50 Sephadex columns in 96-well Millipore filter plates (Multi Screen-HV plates, Cat# MAHVN4510; Millipore, Bedford, MA). For a more extended comparison of read length and signal strengths using many different purification protocols, see reference 7.
5. Lowering the number of cycles needed for optimal results.
   In standard DNA sequencing protocols, 25 to 30 cycling steps are recommended to obtain optimal signal strengths and read lengths. Because of the efficient transformation to ss DNA when a heat-denaturation step is included, the number of cycles can be reduced to 10 to 15, thereby reducing by half the time needed to perform this step (Figure 1-17).
6. The number of base call errors compared to a known reference sequence is substantially lower when a heat-denaturation step is included. For pGem3zf DNA standard this number was $4.4 \pm 1.2/1000$ bases lower compared when no heat denaturation step was part of the protocol. For many difficult templates, though, the result is no good calls to clear calls and the estimates are difficult (7).
7. The first correctly called base is by 3 to 5 bases closer to the 3' end of the primer when heat denaturation step is included (7).

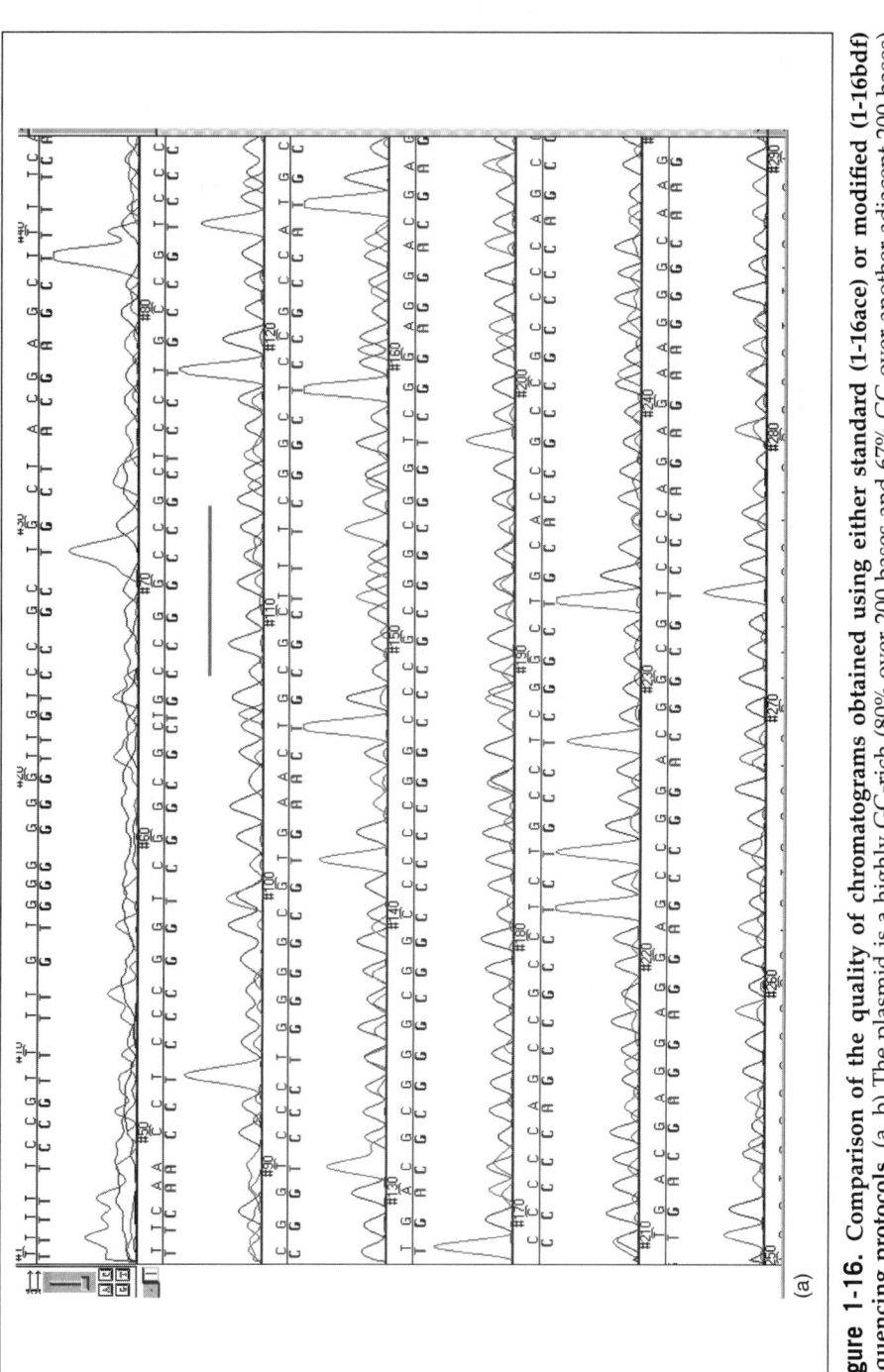

Figure 1-16. Comparison of the quality of chromatograms obtained using either standard (1-16ace) or modified (1-16bdf) sequencing protocols. (a, b) The plasmid is a highly GC-rich (80% over 200 bases and 67% GC over another adjacent 200 bases) DNA template. The Q ≥ 20 values are 486 and 767 bases for standard and modified protocols, respectively.

(continued)

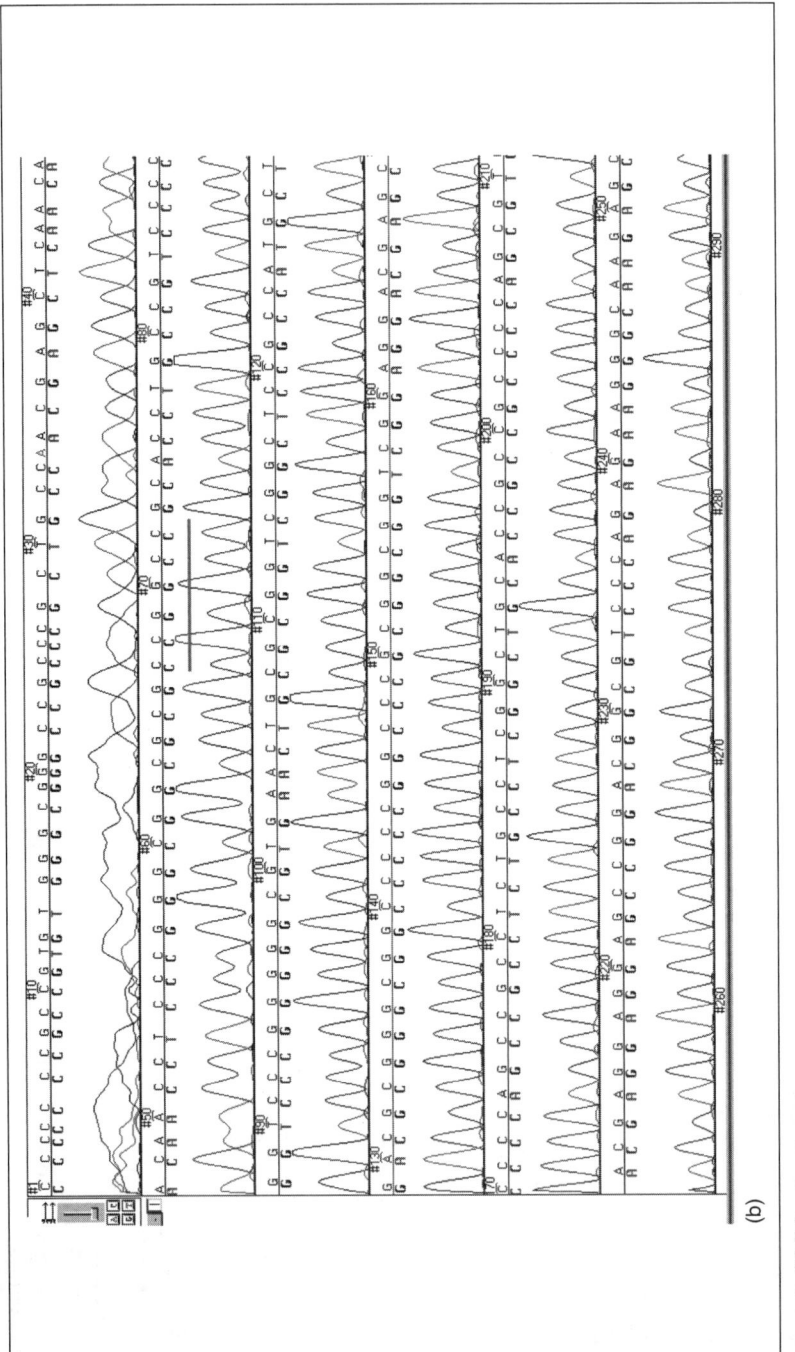

Figure 1-16(b). *Continued*

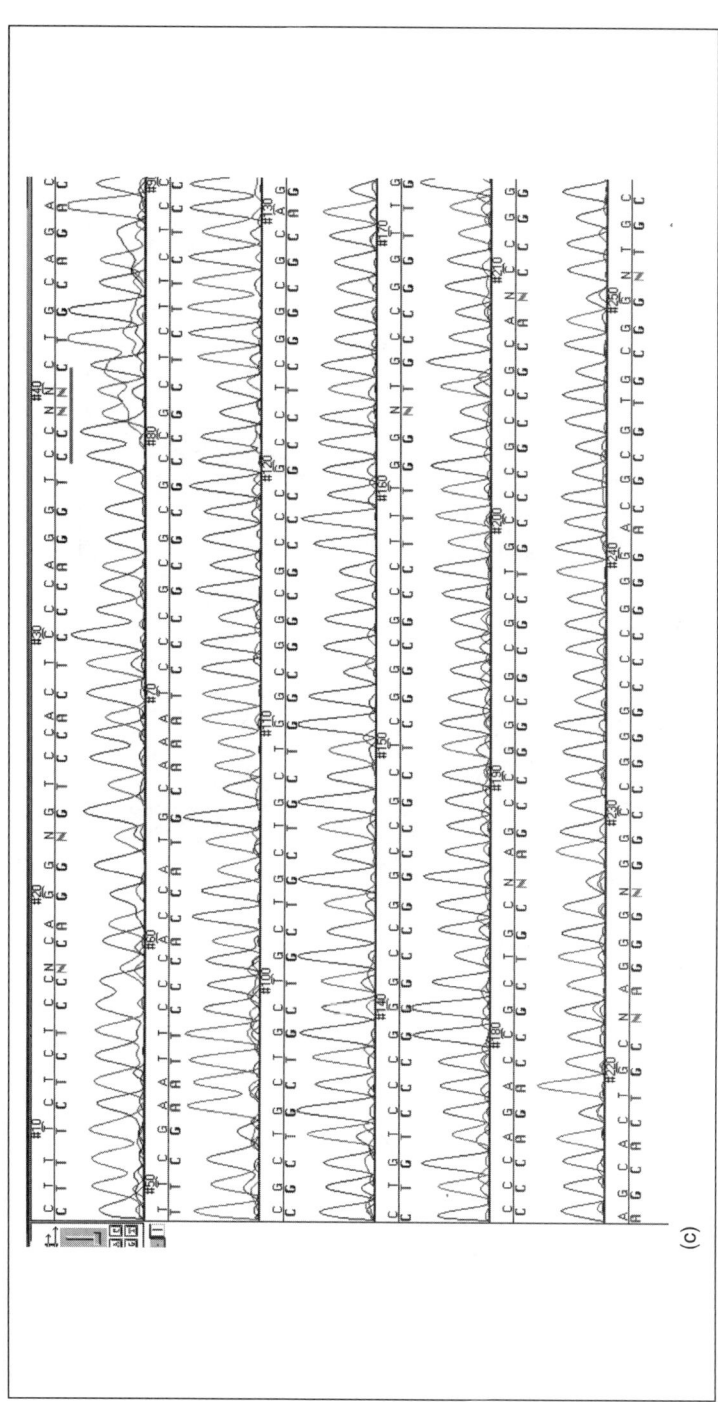

Figure 1-16. (c, d) The plasmid is also a highly GC-rich DNA template: 72%–75% over 600 bases with 200 base region with 80% GC. It also contains 6 CTG repeats. The Q ≥ 20 values are 355 and 864 bases for standard and modified protocols, respectively.

(continued)

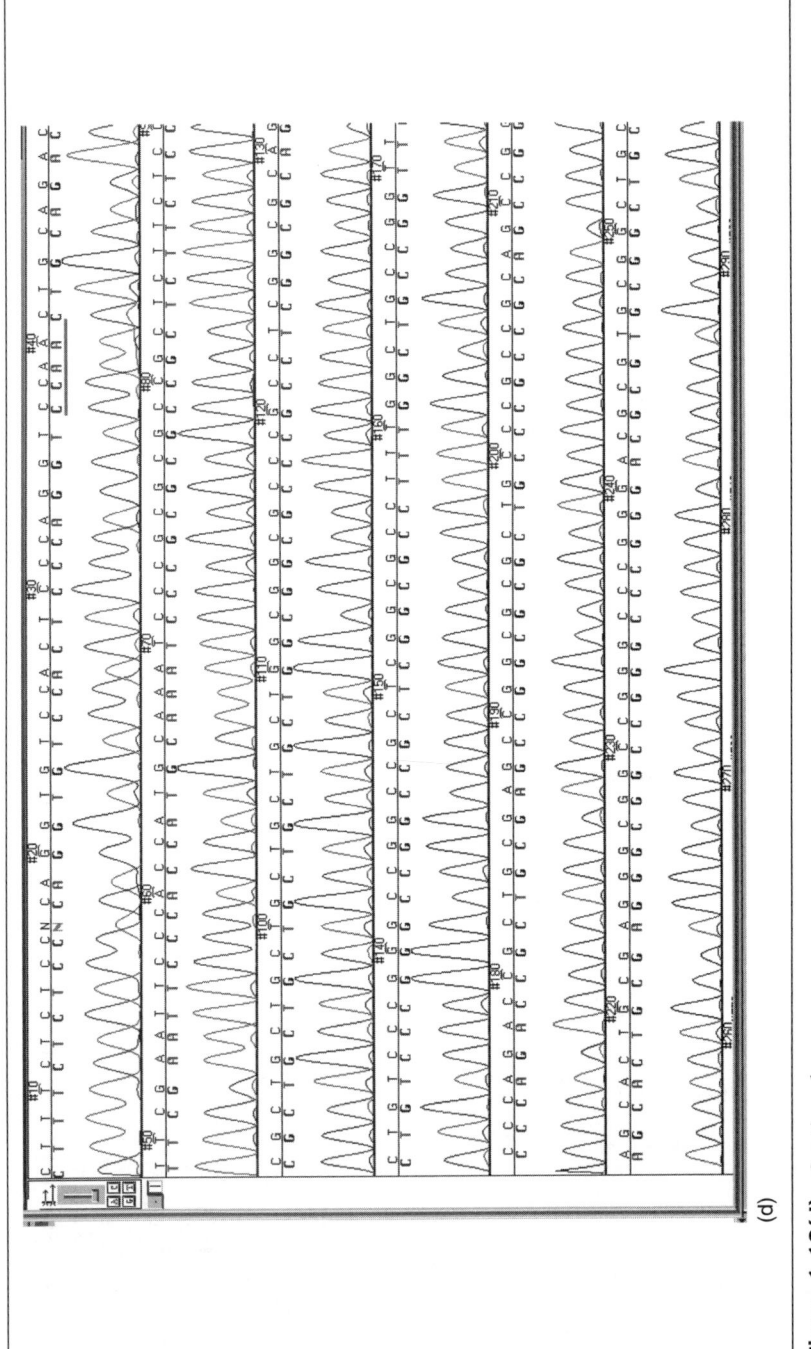

Figure 1-16(d). *Continued*

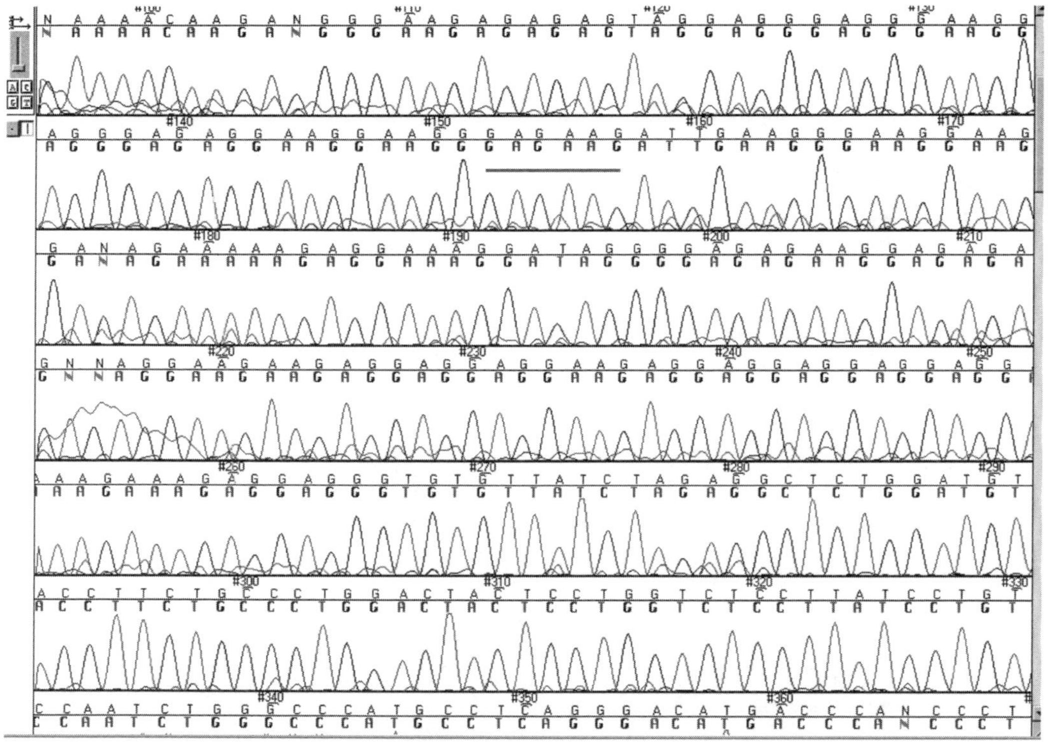

(e)

Figure 1-16(e). (e, f) The plasmid is only about 63% GC-rich in first 200 bases but contains 157 two base (G/A) non-repeats. The Q ≥ 20 values are 566 and 951 bases for standard and modified protocols, respectively. In all of the presented cases, samples that were sequenced using standard protocol are much shorter and have much higher background that leads to more miscalls. The gray bar indicates the same sequences in both paired figures.

(continued)

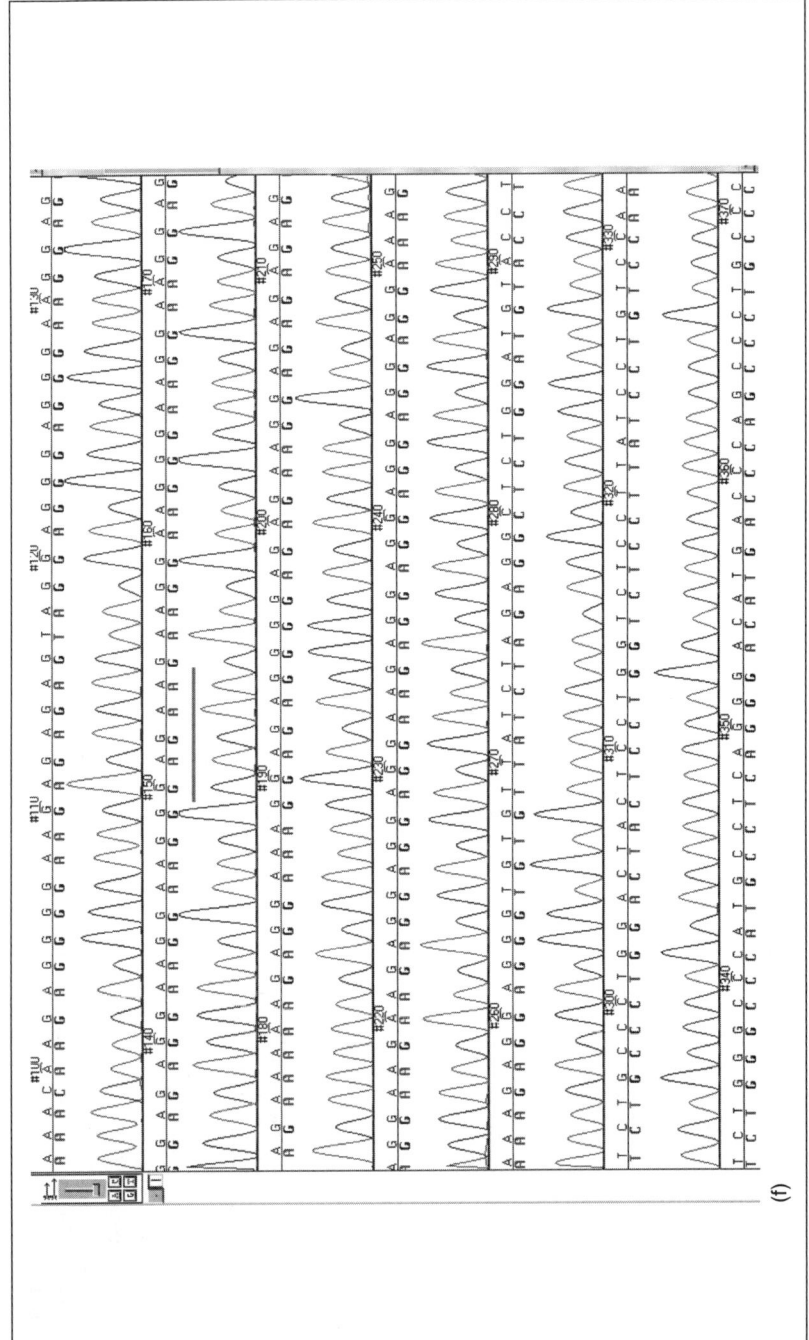

Figure 1-16(f). *Continued*

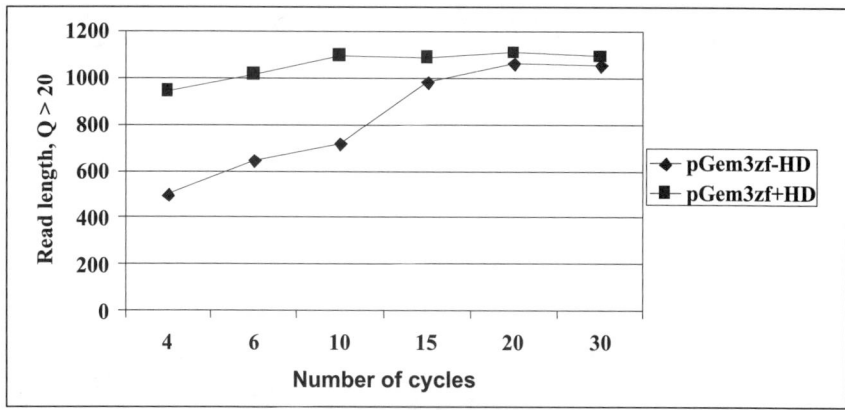

Figure 1-17. **Read length versus number of cycles.** Number of 200 ng aliquots of pGem3zf DNA was subjected to cycle sequencing using either standard (♦; no heat denaturation, –HD) or sequencing modified protocol (■; 5-min of heat denaturation, +HD). As indicated, the number of cycle tubes were withdrawn from the thermocycler and stored at –20°C until the end of the series. Following the cleanup step, samples were run on an ABI3100 DNA analyzer equipped with 80 cm capillary under the conditions suggested by the manufacturer. Read lengths were determined using Q ≥ 20 quality values. Each data point is the average of three reads and standard deviation did not exceed 10% of the average.

Conclusions

The experiments described above led us to introduce an additional step into a standard ABI-like DNA sequencing protocol as shown in Figure 1-18. The adapted protocol not only reduced the amount of template needed to carry out DNA sequencing but also resulted in a greater success rate by eliminating many causes of template "difficulty." For example, before introducing this step, the success rate in our core laboratory was on the order of 65% to 70%, and after the heat-denaturation was implemented, the success rate climbed to 80% to 90%.

Adding this simple heat denaturation step requires less DNA and leads to the increased uniformity, longer read lengths, fewer miscalls, and it is achieved without any additional reagents or capital expenditures. The data also show that PCR fragments of various sizes (at least in a range of 0.6–3.2 kbp) can be treated in the same fashion as plasmid DNAs without the fear of degradation that potentially could lead to lower data quality. This step can be incorporated into any core or high-throughput DNA sequencing laboratory, even if the assembly of sequencing reactions takes few hours. There is no need to take any special precautions as the renaturation to original form is quite slow, thus providing ample time to

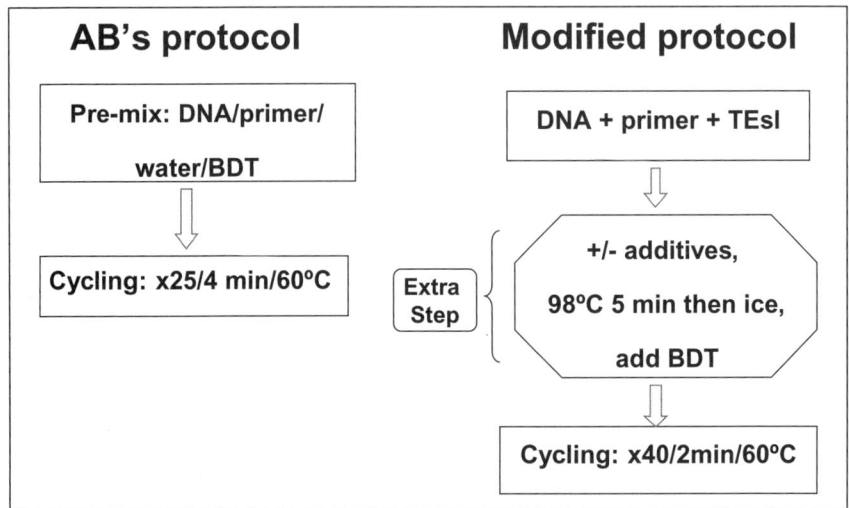

Figure 1-18. Schematic representation of standard and modified DNA sequencing protocols. ABI protocol recommends using 8 μL of Taq mix cocktail, 0.2 to 0.5 μg plasmid DNA, 1 μL of 3.2 mM primer, and water to a total volume of 20 μL. Modified protocol recommended in this work: 2 μL of Taq mix cocktail, 0.1 to 0.5 μg (default = 0.25 μg) of plasmid DNA (or 5–50 ng of PCR fragment; default = 25 ng), 1 μL of 5 mM primer, and TE_{sl} to total volume of 10 μL. Magnesium is adjusted to a final concentration of 2 mM. The ABI's cycling protocol is: 10 seconds at 96°C/5 seconds at 50°C/4 minutes at 60°C. Repeat this cycle 24 times. The modified cycling protocol is: 10 seconds at 96°C/5 seconds at 50°C/2 minutes at 60°C. Repeat this cycle 40 times. The purification of sequencing reactions was carried out as described under DNA sequencing section. The number of cycling steps can be adjusted depending on the particular sequencing machine, clean up method, and the quantity of the starting DNA template used in the sequencing. A good starting point is to use the manufacturer's recommended conditions and then optimize it for a particular laboratory requirement. AB = Applied Biosystems. BDT = Big Dye Terminator. Additives = betaine, reagents A-G, DMSO, etc.

perform any specialized steps one may envision in a typical or specialized sequencing environment.

Acknowledgments

I wish to thank Drs. C. Anderson, J. Dunn, and P. Freimuth of Brookhaven National Laboratory for the gift of some plasmids used in these experiments. I also would like to thank Drs. Alex Ruzin, David Keeney, and Jeff Janso of Wyeth for permission to use their PCR fragments in many experiments presented in this work.

References

1. Bartlett, J.A., Gaillard, R.K., and Joklik, W.K. 1986. Sequencing of supercoiled plasmid DNA. *BioTechniques* 4: 208–209.
2. Friedberg, E.C., Walker, G.C., and Siede, W. 1995. *DNA Repair and Mutagenesis.* Washington, D.C.: W.H. ASM Press.
3. Haltiner, M., Kempe, T., and Tjian, R. 1985. A novel strategy for constructing point mutations. *Nucleic Acids Res* 13: 1015–1025.
4. Hattori, M., and Sakaki, Y. 1986. Dideoxy sequencing method using denatured plasmid templates. *Anal Biochem* 152: 232–238.
5. Holmes, D.S., and Quiglev, M. 1981. A rapid method for the preparation of plasmids. *Anal Biochem* 114: 193–197.
6. Kieleczawa, J. Controlled heat-denaturation of DNA plasmids. In: Kieleczawa, J. (ed). *DNA Sequencing: Optimizing the Process and Analysis.* Sudbury, MA: Jones and Bartlett; 2005: 1–10.
7. Kieleczawa, J., and Bajson, K. Evaluation of methods for cleanup of DNA sequencing reactions. In: Kieleczawa, J. (ed). *DNA Sequencing II: Optimizing Preparation and Cleanup.* Sudbury, MA: Jones and Bartlett; 2006: 219–240.
8. Kieleczawa, J., and Wu, P. 2006. Prolonged storage of plasmid DNAs under different conditions: effects on plasmid integrity, spectral characteristics, and DNA sequence quality. In Kieleczawa, J. (ed). *DNA Sequencing II: Optimizing Preparation and Cleanup.* Sudbury, MA: Jones and Bartlett; 2006: 259–274.
9. Lee, J-S. 1991. Alternative dideoxy sequencing of double-stranded DNA by cyclic reactions using Taq polymerase. *DNA Cell Biol* 10: 67–73.
10. Murray, V. 1989. Improved double-stranded DNA sequencing using the linear polymerase chain reaction. *Nucleic Acids Res* 17: 8889.
11. Sambrook, J., and Russell, D.W. *Molecular Cloning,* 3rd ed. Cold Spring Harbor, NY: CSH Laboratory Press; 2001.
12. Sauer, P., Muller, M., and Kang, J. 1998. Quantitation of DNA. *Qiagen News* 2: 23–26.
13. Toneguzzo, F., Glynn, S., Levi, E., et al. 1988. Use of a chemically modified T7 DNA polymerase for manual and automated sequencing of supercoiled DNA. *BioTechniques* 6: 460–469.
14. Wilfinger, W.W., Mackey, K., and Chomczynski, P. 1997. Effect of pH and ionic strength on the spectroscopic assessment of nucleic acid purity. *BioTechniques* 22: 474–481.
15. Wilfinger, W.W., Mackey, K., and Chomczynski, P. Assessing the quality, purity and integrity of RNA and DNA following nucleic acid purification. In: Kieleczawa, J. (ed). *DNA Sequencing II: Optimizing Preparation and Cleanup.* Sudbury, MA: Jones and Bartlett; 2006: 291–312.
16. Yie, Y., Wei, Z., and Tien, P. 1993. A simplified and reliable protocol for plasmid DNA sequencing: fast miniprep and denaturation. *Nucleic Acids Res* 21: 361.

2 Heat Denaturation Is an Effective Step in Sequencing of Many Difficult DNA Templates

Jan Kieleczawa
Wyeth Research, Cambridge, MA

The last three decades have witnessed amazing progress in the development of many tools and techniques in molecular biology, one of which is DNA sequencing. Since its humble beginnings (23, 25) when it still took a considerable amount of skill and effort to sequence even a relatively small molecule, DNA sequencing is now almost as routine as, for example, gene cloning or polymerase chain reaction (PCR) technology. Quite often, the only barrier to performing any sequencing experiment is the cost of a sequencing instrument itself (which ranges from below $100,000 to about $500,000 depending on the manufacturer and required throughput). In the simplest incarnation of Sanger sequencing (25), all one needs to do is to combine a few components (DNA, primer, water, additives, if any, and detection mix), and then perform cycle sequencing, purify, and run on a detection instrument. If all this is carried out correctly, one can expect up to 1000 readable bases/reaction. This is true most of the time, but only when the starting DNA template is of good quality, sufficient quantity, and does not contain regions that are considered difficult. For the purposes of this chapter, we consider a template (or just some part of it) to be difficult when it cannot be sequenced using a standard ABI-like DNA sequencing protocol (1). Difficult templates can be classified into the following categories:

a. *GC-rich:* The lower threshold percentage of GC at which a template is considered difficult is not commonly accepted, although a template (or at least 100–150 base long part of it) with >60% to 65% GC content could reasonably be considered difficult (2–4, 15).

DNA III: Dealing with Difficult Templates
Edited by Jan Kieleczawa
©2008 Jones and Bartlett Publishers

b. *Containing various repeats:* These include: di- and trinucleotide, direct, inverted, and Alu repeats. Examples are: AG, CA, CT, GT, AGG, ACC, CCG, CCT, CTT, GCC, GGA, CCCTTT, and any other combination of these repeats (15, 35). From a sequencing point of view, long stretches (>40) containing only two nucleotides that do not necessarily form repeats also can be difficult to sequence (this work).

c. *Containing hairpin structures:* Technically, these structures consist of two inverted repeats, separated by at least three nucleotides. They warrant classification as a separate category due to the increasing importance of si/shRNA-based research that quite often requires sequencing through such structures (5, 8, 13, 15, 16, 27). Strong hairpin structures are an integral part of some vectors, for example DONR/pDEST series from Invitrogen (www.invitrogen.com) or as in Inverted Terminal Repeat (ITR) sequences in adeno-associated viruses.

d. *Containing long homopolymer stretches*: Most typical are poly A/T tails resulting from constructs obtained through reverse transcriptase (RT)-PCR amplification of mRNAs. The length of such tails varies from around 20 to over 100 A/Ts. The poly G/C stretches not only contributes to the overall GC richness but also may form strong hairpins and other complex secondary structures.

e. *Containing sequence motifs causing band compressions* (33): These are predominantly the following motifs: 5'-YGN$_{1-2}$AR where Y and R are pyrimidine and purine residues, respectively, and N can be any base.

Chapter 5 has further information about the nature of these difficult templates.

One of the complicating factors in sequencing many different categories of difficult templates is the increasingly apparent realization that there may not be a "one-method-fits-all" solution and that each category requires a separate approach (or even set of approaches). A number of articles describe modified sequencing protocols. However, at best they appear to apply only to specific types of difficult templates (2–4, 7, 9, 11, 15, 22, 31).

On the other hand, a single approach that incorporates a five-minute heat denaturation step (with or without some other additives) was successfully applied to many different categories of difficult templates (15–17, 21).

This chapter reviews both general and specific approaches to sequencing many different kinds of difficult templates, starting with the description of the heat denaturation step and its impact on the quality of sequencing data.

Materials and Methods

All materials and methods presented in this review were extensively described in earlier publications (15–21) and will be not repeated here unless there is a notable deviation. However, for convenience, we will define "standard" and "modified" sequencing protocols. *Standard sequencing protocol* refers to a basic protocol recommended by Applied Biosystems technical literature as described (e.g., in reference 3), with the exception that the final volume is 10 μL. Briefly, DNA, primer, water, and dye terminator mix (V 3 or 3.1; purchased from Applied Biosystems, Inc., Foster City, CA) are combined and cycled 25 times [(96°C/10 seconds) (50°C/5 seconds) (60°C/4 minutes)]. In the *modified sequencing protocol*, DNA, primer, and 10 mM Tris pH 8.0/0.01 mM EDTA (TE$_{sl}$) are combined; the samples are heat denatured for 5 minutes at 98°C and then added to the dye-terminator mix. If additives are used, they are included in the heat denaturation step. Following the cycling step, reactions were purified using Performa® V3 96-Well Short Plate (EdgeBiosystems, Gaithersburg, MD) or using in-house prepared 96-well plates filled in with G50 Sephadex beads. The 96-well plates were purchased from Innovative Microplate (catalog #F20011; Chicopee, MA) and superfine Sephadex was from GE Healthcare (catalog #17-0041-01; Fairfield, CT). Purified samples were run on ABI's genetic analyzer (throughout these studies ABI377, 3700, 3100, and 3730 were used) under the conditions recommended by the manufacturer, specific for each instrument. Genetic analyzers were purchased from Applied Biosystems, Inc.

The following DNA polymerases were used to amplify difficult regions: Go Taq® DNA Polymerase (Promega, Madison, WI); KOD HiFi DNA Polymerase (Novagen, La Jolla, CA); Phusion™ DNA Polymerase (BioRad, Hercules, CA); TopoTaq DNA polymerase (Fidelity Systems, Inc., Gaithersburg, MD); Platinum® Taq DNA Polymerase High Fidelity (Invitrogen, Carlsbad, CA); and Vent® DNA Polymerase (New England Biolabs, Beverly, MA). We followed the manufacturers' recommendations specific to each enzyme for amplification of desired fragments. In each case though, dGTP nucleotide was substituted with 7-deaza-dGTP (all nucleotides were from GE Healthcare). In addition, all PCR fragments were amplified under the standard PCR conditions (with all 4 regular nucleotides). The PCR products were purified using QuickStep™ 2 PCR Purification Kit (EdgeBiosystems) and PCR products were sequenced using standard sequencing protocol as described above. For controls, the DNA sequencing also was carried out on parent plasmids that were used in obtaining PCR fragments. Following the PCR cleanup step, a 5 μL aliquot of each amplicon was run on an E-gel agarose system (Invitrogen) under conditions recommended by the manufacturer. Low DNA Mass Ladder (cat #10068-013; Invitrogen) was used to estimate the concentration of each PCR amplicon. Note that the intensity of bands amplified with 7-deaza-dGTP is lower compared to control bands obtained under standard conditions, and visual inspection greatly underestimates the DNA concentration (15 and references therein).

Primer on Controlled Heat Denaturation of Plasmids

An extensive description of the heat denaturation process is given in Chapter 1.

In the early days, DNA sequencing was dominated by non-thermo-stable polymerases, primarily the T7-based Sequenase® (28–30), and to sequence any double-stranded template, the first step needed to be strand separation for the efficient annealing of the primer. There were many protocols designed for strand separation, all including heating (22° to 100°C) in the presence or absence of 0.1 to 0.3 N NaOH, followed by neu-tralization, precipitation and re-suspension in the desired solution (12, 14, 32, 34). These protocols were time-consuming, cumbersome, and insuffi-ciently reliable for routine use. Instead, a simple heat denaturation step was suggested in low-salt buffers and at elevated temperatures to convert supercoiled plasmid DNA efficiently to a single-stranded form amenable to sequencing (17). The denaturation can be carried out in water, but it occasionally produces additional bands that effectively reduce the amount of template available for sequencing (J. K., unpublished observation).

The time needed to convert DNA effectively from supercoiled form to single-stranded form (ss form) depends on the size of the plasmid; the bigger the plasmid, the shorter time needed for this conversion. It takes 7.5 minutes (10 mM Tris-Cl, pH 8.0 buffer/98°C) to convert 75% of pGem3zf (3.2 kbp) from supercoiled to ss form. (The 75% conversion level is selected to avoid potential degradation of DNA.) For any plasmid bigger than pGem3zf, one needs to subtract 1 minute per multiple of 2.5 kbp from 7.5 minutes to achieve a similar level of denaturation. In fact, there is a linear relationship between the size of a plasmid, at least in the 3 to 20 kbp size range, and the time needed for efficient conversion to a form suitable for sequencing (17). As a rule of thumb, the time needed to achieve a similar level of denaturation in water should be halved compared to the time in 10 mM Tris-Cl buffer. Furthermore, the linear relationship between the size of the plasmid and the time needed for conversion only holds true for plasmids that do not contain any difficult regions. For example, templates rich in GC or with CTT require 30 and 20 minutes, respectively, for 75% conversion. Surprisingly, plasmids with long poly A/T tracts (70–80 bases) also require up to 20 minutes for effective conversion to ss form under similar conditions. In the presence of 2 mM $MgCl_2$, which is the final con-centration of Mg ions under optimal cycle sequencing conditions using

ABI's dye-terminator chemistry (3), there is no, or very little, conversion of supercoiled to ss form either during prolonged heat denaturation at 98°C or during cycle sequencing (see Chapter 1). It appears the primary part that is transformed to a form amenable for sequencing is the nicked form; hence one needs to use more DNA than necessary to compensate for the partial conversion. Based on the data presented (15, 17, 18), the following are advantages to incorporation of the heat denaturation step into a sequencing protocol:

a. Very effective conversion of supercoiled DNA to ss form and destruction of any residual DNAse activity, if present.

b. The amount of DNA needed for optimal results is three- to fivefold lower than that recommended by a standard ABI protocol.

c. At a given amount of DNA, the fluorescent signal strength is approximately threefold higher compared to samples without a heat denaturation step (which is directly related to point a above).

d. Increased read length ($Q \geq 20$) by 72.3 ± 20 (~10% more) for standard (non-difficult) templates like pGem3zf.

e. Very effective when sequencing many types of difficult templates. In all cases, clearer data are generated with this step; in 10 of 57 cases, the heat denaturation method succeeded in generating readable data where standard methods failed (Table 2-1).

f. First correctly called base is three to five nucleotides closer to the 3' end of the primer (for M13 primer with pGem3zf DNA).

g. Fewer sequencing errors detected; 4.4 errors/1000 bases less compared to a standard protocol (again this is for a standard, non-difficult template).

The kinetics of reversal for the ss form back to a supercoiled-like form is extremely slow (measured in days) both in the presence or absence of 2 mM MgCl$_2$ and at temperatures ranging from 0° (ice) to 22°C (see Chapter 1). So even if the assembly of sequencing reactions takes few hours—as it could in a factory-like sequencing pipeline—there is no need to take any special precautions between the initial heat denaturation (performed with DNA, primer, low-salt buffer, or water), and before the addition of dye-terminator mix. Furthermore, as indicated by Kieleczawa and Wu (19), the dye-terminator mix can be stored for several days at room temperature (or even at 37°C) with no adverse influence on the read length and the quality of the resulting sequences. Hence, all assembly and pretreatment steps can be safely performed at room temperature, with the only necessary precaution being control of evaporation of the sequencing mix to avoid excessive imbalance in the dNTPs and Mg ion concentrations.

Figures 2-1 to 2-6 (and Plates 2 and 3 in the Color Addendum) provide examples of the effect of heat denaturation on the sequence quality.

Table 2-1. Sequencing difficult templates using various protocols.
Fifty-nine different difficult DNAs were sequenced using variety of protocols. Method 1 is equivalent to the standard ABI protocol (1). Method 2 is equivalent to protocol described in references 7, 13, 20 (includes 5 minutes heat-denaturation step). Method 3 is like protocol 1 but in the presence of 5% DMSO. Method 4 is as protocol 2 but DMSO was included during heat-denaturation step. Method 5 is as protocol 2 but one of seven reagents from Invitrogen's (Carlsbad, CA) sequence enhancer Rx set was present during denaturation step (only data for the longest reads are reported here). Method 6 is like method 2 but included some other additive (GC melt is from BD Biosciences-San Jose, CA; Betaine was from Sigma-Aldrich-St. Louis, MI). Method 7 was like method 2 but BigDye™ V3.1 was substituted with dGTP V3.0 and method 8 was like method 2 but BigDye™ V3.1 was substituted with 4:1 (v/v) mix of BigDye™ V3.1 and dGTP V3.0 as suggested by G. Grills (personal communication). All data are expressed in Q ≥ 20 values.

		Method							
	Sequencing → Protocol	#1	#2	#3	#4	#5	#6	#7	#8
DNA #	DNA characteristics ↓	−HD	+HD	−HD	+HD	+HD	+HD	+HD	+HD
1	60% GC over 200 bases	834	868	818	862	895 ACFG	884 Bet	511	280
2	62% GC over 800 bases	849	890	861	904	913 ACFG	940 Bet	688	859
3	64% GC over 800 bases	0	830	NT	853	929 C	893 G-C	855	NT
4	66% GC over 800 bases	538	847	867	854	894 AG	858 Bet	728	635
5	68% GC over 400 bases	0	989	NT	222	1087 C	1025 G-C	NT	181
6	69% GC over 630 bases (366 Gc/70 Cs): F	110	665	NT	NT	732 CF	597 Bet	0	0
7	69% GC over 630 bases (248 Cs/56Gs): R	0	133	100	122	244 CG	248 Bet	0	0
8	70% GC over 800 bases	661	920	914	920	940 ACG	939 Bet	858	848
9	70% GC over 300 bases	0	989	NT	222	1087 C	1025 G-C	177	945

#	Description								
10	72% GC over 150bp/65% over 300 bases	208	572	NT	121	673 C	525 G-C	562	549
11	75% GC over 600 bases	0	660	NT	960	974 F	940 Bet	747	713
12	78% GC over 150 bases	0	504	NT	928	967 C	988 G-C	749	675
13	78% GC over 400 bases	390	609	503	805	960 AF	1000 Bet	905	890
14	80% GC over 200bp/67%GC over next 200 bases	334	682	NT	NT	769 A	370 Bet	565	NT
15	80% GC over 400 bases	378	991	873	992	996 ACG	995 B	882	886
16	90% GC over 200 bases/73% GC over next 400 bases	338	705	607	587	688 C	628 Bet	478	693
17	94% GC over 200 bases/101 base non-repeat G/C*	0	277	NT	270	304 A	302 B	318	420
18	65% GC over 200 bases/18 × CGC repeat	585	668	613	608	833 F	595 Bet	256	184
19	72% GC over 200 bases/6 × GCT repeat	495	515	472	404	816 A	806 Bet	668	723
20	75% GC over 600 base/7 × CTG repeat: F	250	291	328	354	770 A	845 Bet	475	169
21	75% GC over 600 bases/7 × GAC repeat: R	780	843	766	839	790 A	815 Bet	733	765
22	12Cs homopolymer/another 10 Cs 19 bases apart	728	922	885	921	999 AC	951 Bet	NT	NT
23	10Gs homopolymer/19Gs 26 bases apart: F	706	743	809	834	860 AC	840 Bet	475	NT
24	19Cs homopolymer/10Cs 26 bases apart: R	487	537	403	409	856 AC	750 Bet	471	NT

(continued)

Table 2-1. *Continued*

DNA #	Sequencing → Protocol / DNA characteristics ↓	Method #1 -HD	#2 +HD	#3 -HD	#4 +HD	#5 +HD	#6 +HD	#7 +HD	#8 +HD
25	18 Cs/10 Cs separated by 7 bases: F	332	435	161	115	519 C	628 Bet	95	141
26	18 Gs/10 Gs separated by 7 bases: R	0	99	0	0	0	826 Bet	413	389
27	27 Gs followed by 23 G/C non-repeat	243	472	238	378	553 A	540 Bet	232	NT
	15bp inverted repeat with 73 base loop 6 × GGC 75% GC over 600 bases								
28	37Gs homopolymer followed by 27bp G/C non-repeat. Also 15bp inverted repeat	283	404	407	418	606 A	NT	605	NT
29	33 bases of poly A in the middle of a trace	885	899	863	850	940 ACG	931 Bet	547	408
30	51 base non-repeat C/T	222	277	185	193	914 CD	898 Bet	477	NT
31	65 base non-repeat G/A	823	862	528	668	550 C/G	466 Bet	327	647
32	65 base non-repeat T/C/67% GC over 200 bases: F	387	597	NT	NT	940 C	NT	247	NT
33	65 base non-repeat A/G + 64% GC over 400 bases: R	758	917	NT	NT	950 C	NT	537	NT
34	68 base non-repeat A/G (27As followed by 41 non-repeat A/G)	822	921	NT	NT	895 A	896 Bet	884	NT
35	101 base non-repeat C/T	283	409	272	335	445 F	298 Bet	199	220
36	147 base non-repeat G/A: F	363	761	168	416	900 ACG	904 G-C	580	765

37	147 base non-repeat T/C: R	569	886	107	618	950 C	815 Bet	845	927
38	456 base non-repeat G/A*: F	490	504	480	480	478 ACG	484 Bet	553	509
39	456 base non-repeat T/C*: R	401	431	NT	411	540 BCE	466 Bet	455	426
40	30 × CT and 6 × CAT repeats: F	312	328	272	285	622 A	713 Bet	160	197
41	30 × GA and 6 × GTA repeats: R	725	732	446	415	659 C	811 Bet	169	639
42	11 base inverted repeat with 3 base loop	311	356	298	338	382 A	369 Bet	432	NT
43	6 × GCC and 76% GC over 600 bases Alu repeat + 22 base inverted repeat/84 bases loop	0	118	109	107	262 DE	258 G-C	820	NT
44	44 base inverted repeat, series of palindromes, longest 50 bases: F	348	624	881	783	875 A	849 Bet	239	842
45	44 bases inverted repeat, series of palindromes, longest 50 bases: R	580	606	853	771	660 ACG	840 Bet	195	822
46	88 bases inverted repeat/60 bases loop	317	368	477	564	669 F	699 Bet	632	770
47	19 and 15 bases inverted repeats followed by 19 Cs and 41 base non-repeat T/A: F	355	358	345	356	652 A	558 Bet	623	833
48	19 and 15 bases inverted repeats followed by 19 Gs and 41 base non-repeat A/T: R	586	592	583	592	602 A	598 Bet	813	932
49	111 bases inverted repeat separated by 1705 bases: F	120	325	115	351	348 C	336 Bet	783	933

(continued)

Table 2-1. *Continued*

DNA #	Sequencing → Protocol / DNA characteristics ↓	#1 −HD	#2 +HD	#3 −HD	#4 +HD	#5 +HD	#6 +HD	#7 +HD	#8 +HD
						Method			
50	111 bases inverted repeat separated by 1705 bases: R	400	605	291	478	493 C	561 Bet	833	700
51	111 bases inverted repeat separated by 1705 bases: F	235	231	213	228	217 C	218 Bet	870	911
52	111 bases inverted repeat separated by 1705 bases: R	819	918	864	894	945 CG	858 Bet	522	290
53	19 base hairpin: F	348	358	349	356	635 (A)	354 Bet	1035	1027
54	19 base hairpin: R	170	189	175	192	183 (C)	287 Bet	1025	1014
55	21 base hairpin with 2 base bulge: F	534	716	655	712	980 A	940 Bet	580	870
56	21 base hairpin with 2 base bulge: R	792	829	865	912	924 CG	990 Bet	580	870
57	28 base ITR hairpin*	410	439	NT	NT	472 C	462 Bet	NT	NT
58	29 base hairpin with 4 internal mismatches: F	0	590	NT	601	710 ACD	885 Bet	811	811
59	29 base hairpin with 4 internal mismatches: R	566	820	NT	750	806 CDF	806 G-C	963	955

* To get through difficult regions in these three DNAs, we had to apply 2-step protocol using 7-deaza-dGTP PCR (15) or T29 amplification (26).

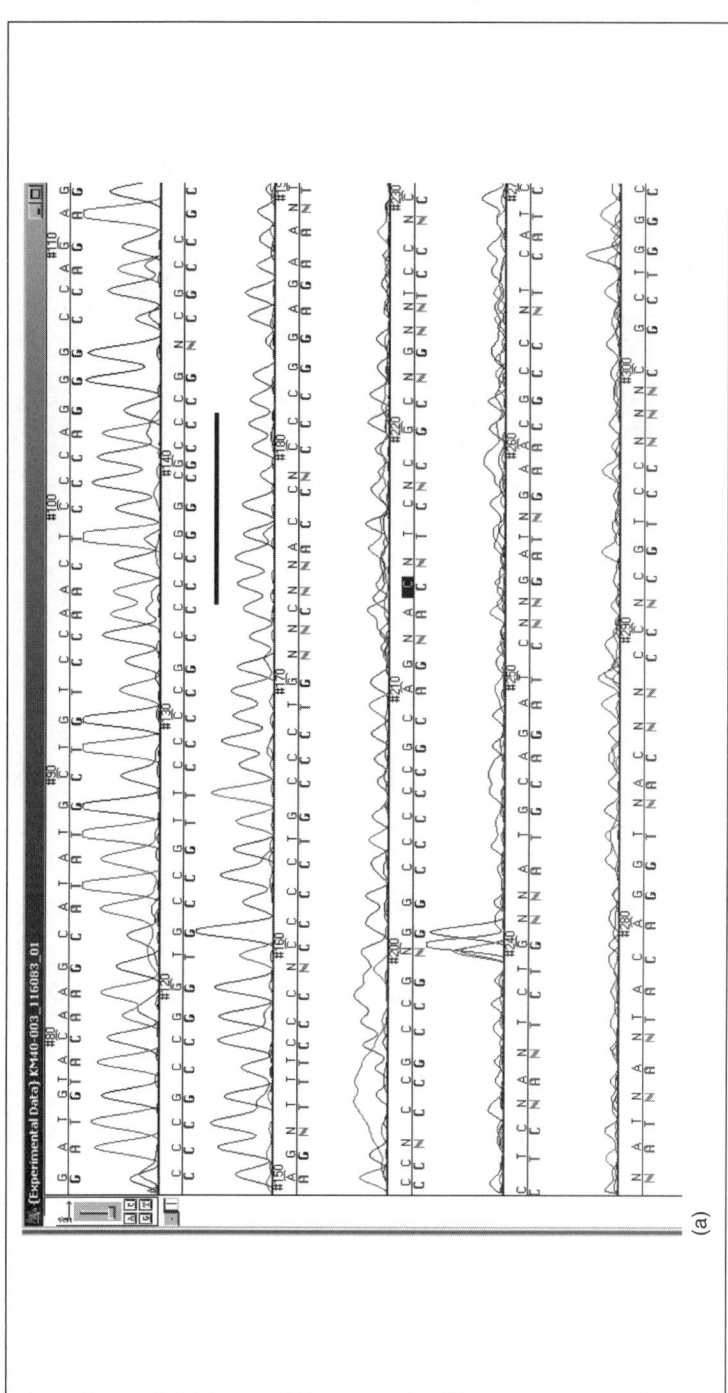

Figure 2-1. Chromatograms of GC-rich template (about 67%) sequenced using a standard (a) and modified (b) protocol. The black bar indicates identical sequence regions.

(continued)

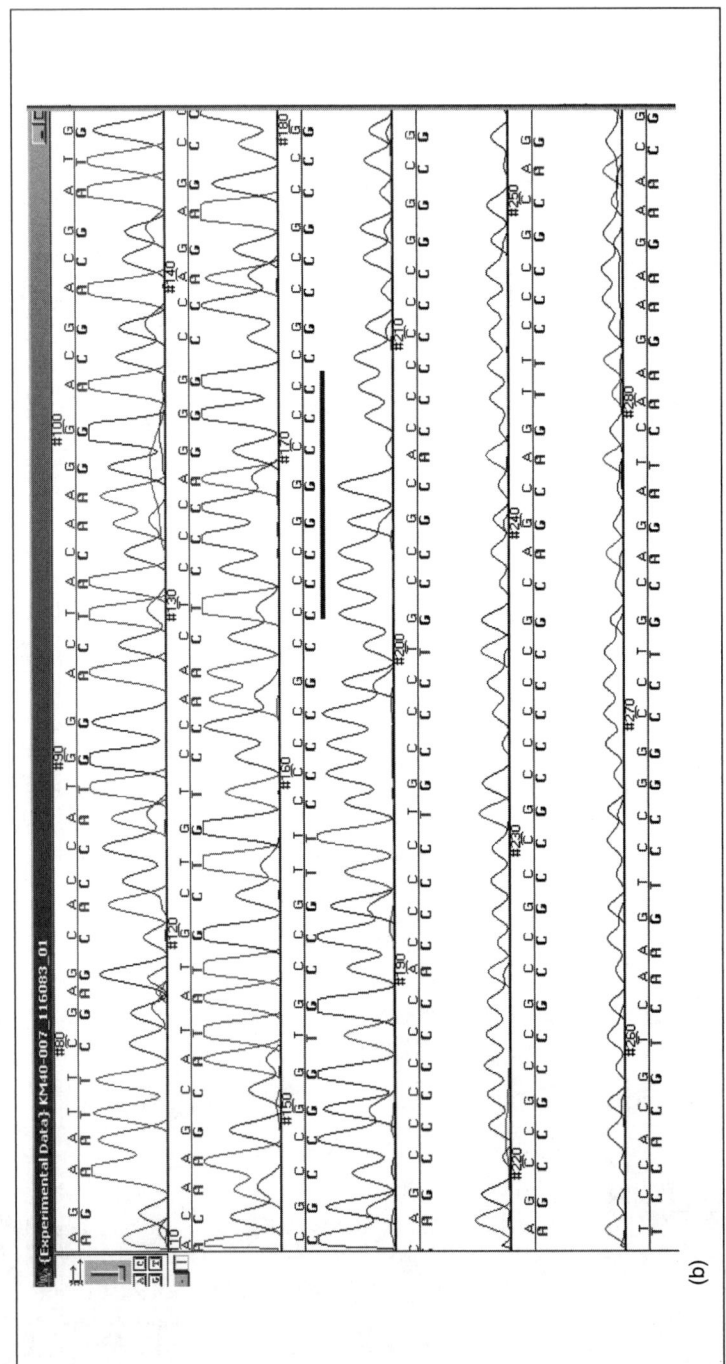

Figure 2-1(b). *Continued*

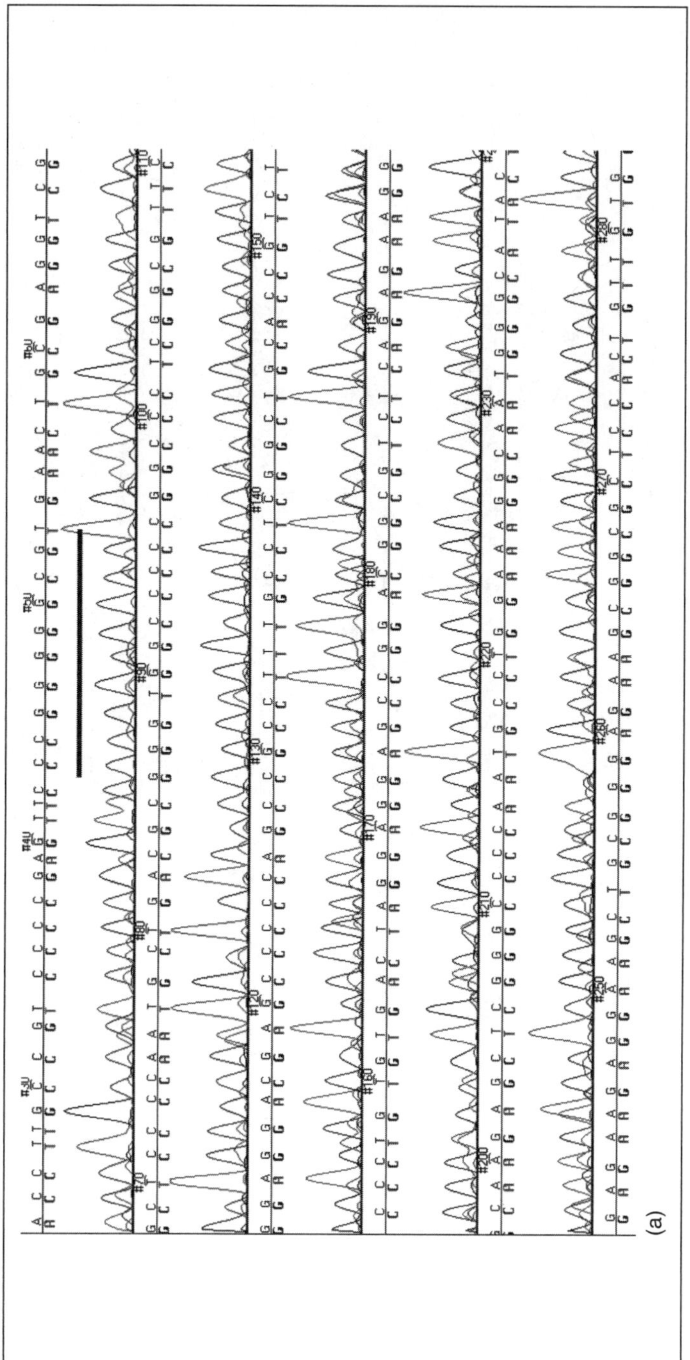

(a)

Figure 2-2. Chromatograms of GC-rich template (about 72%) sequenced using a standard (a) and modified (b) protocol. The black bar indicates identical sequence regions.

(*continued*)

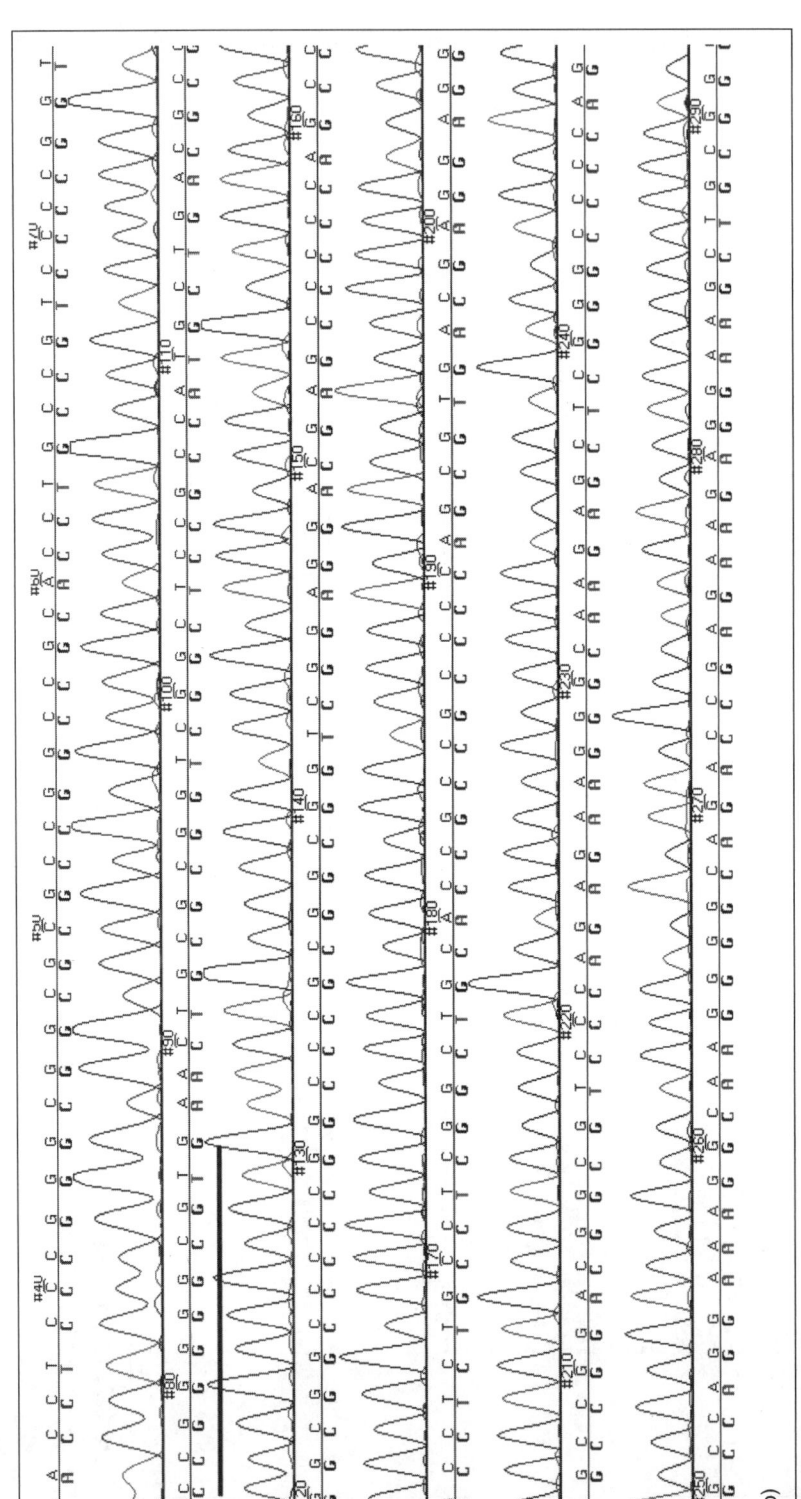

Figure 2-2(b). *Continued*

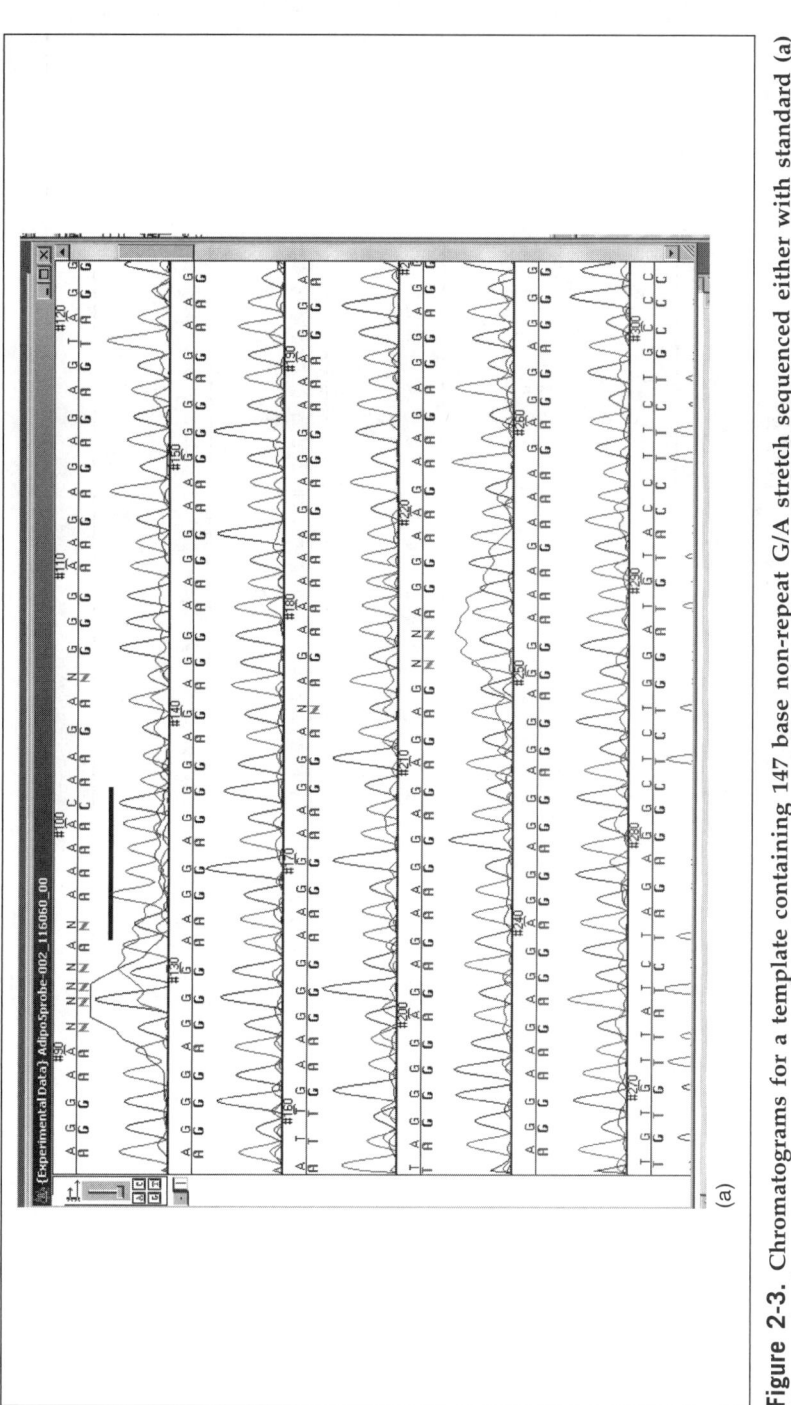

Figure 2-3. Chromatograms for a template containing 147 base non-repeat G/A stretch sequenced either with standard (a) (Plate 2 in the Color Addendum) or modified (b) protocol. The black bar indicates identical sequence regions.

(continued)

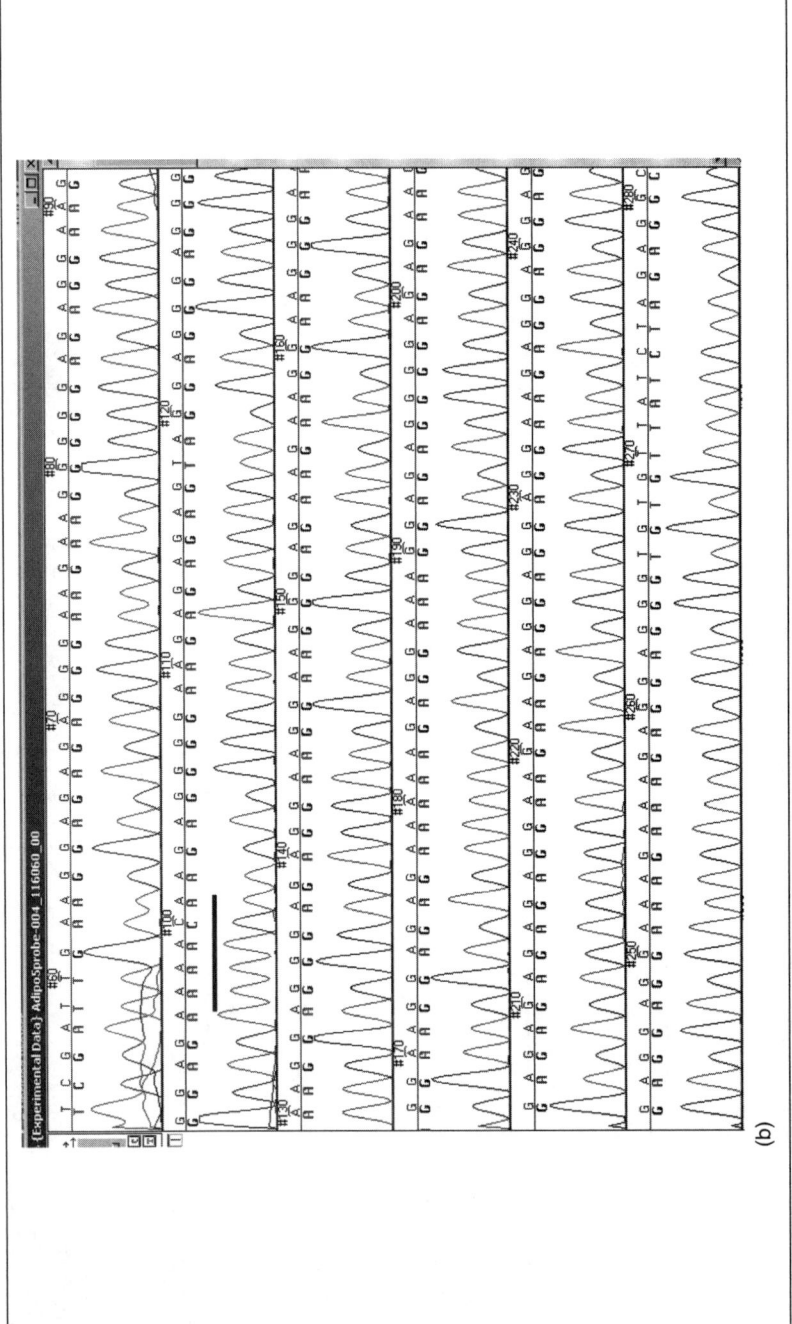

Figure 2-3(b). *Continued*

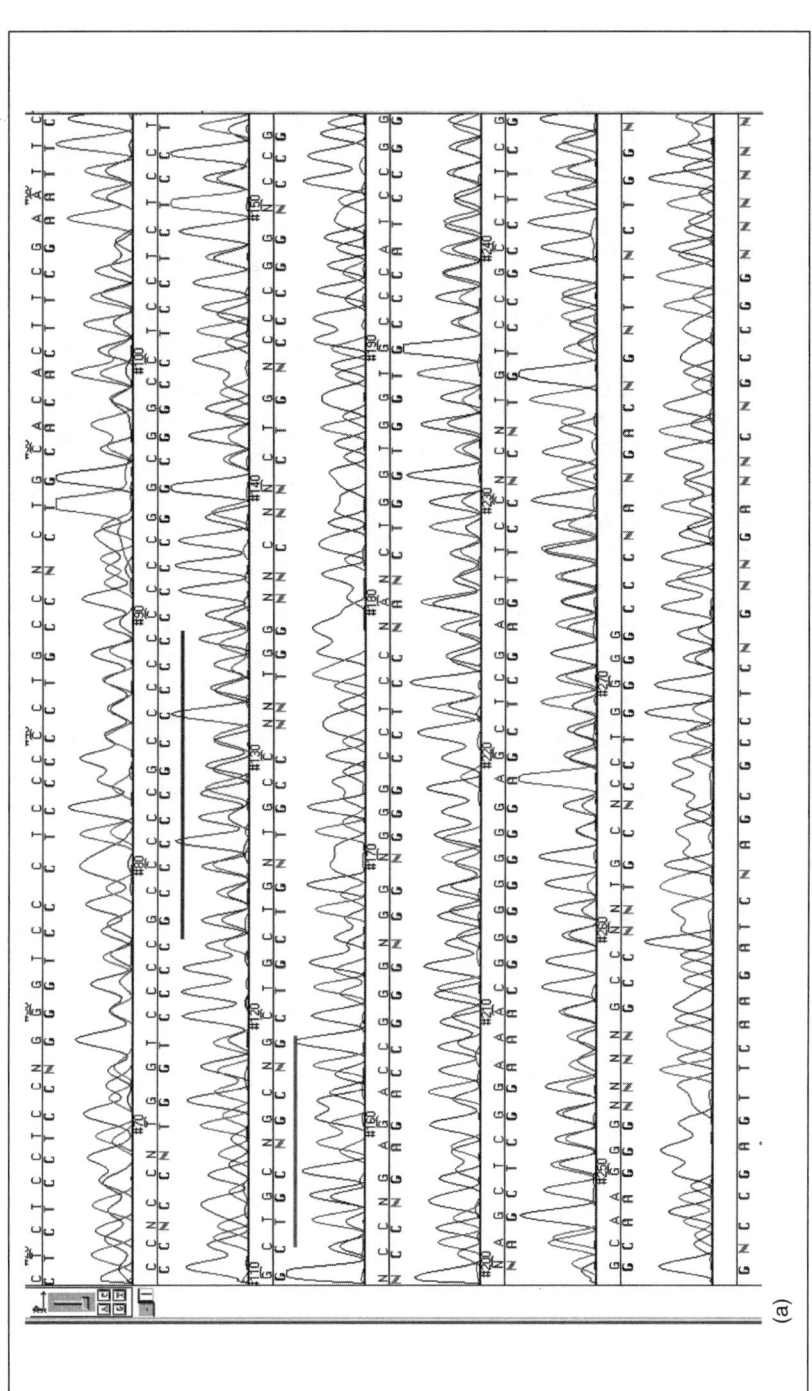

Figure 2-4. An example of chromatograms for a template with C-stretch (indicated with dark gray line) and 7× CTG repeat (indicated with light gray line) sequenced either with standard (a) or modified (b) protocols. The bars indicate identical sequence regions.

(continued)

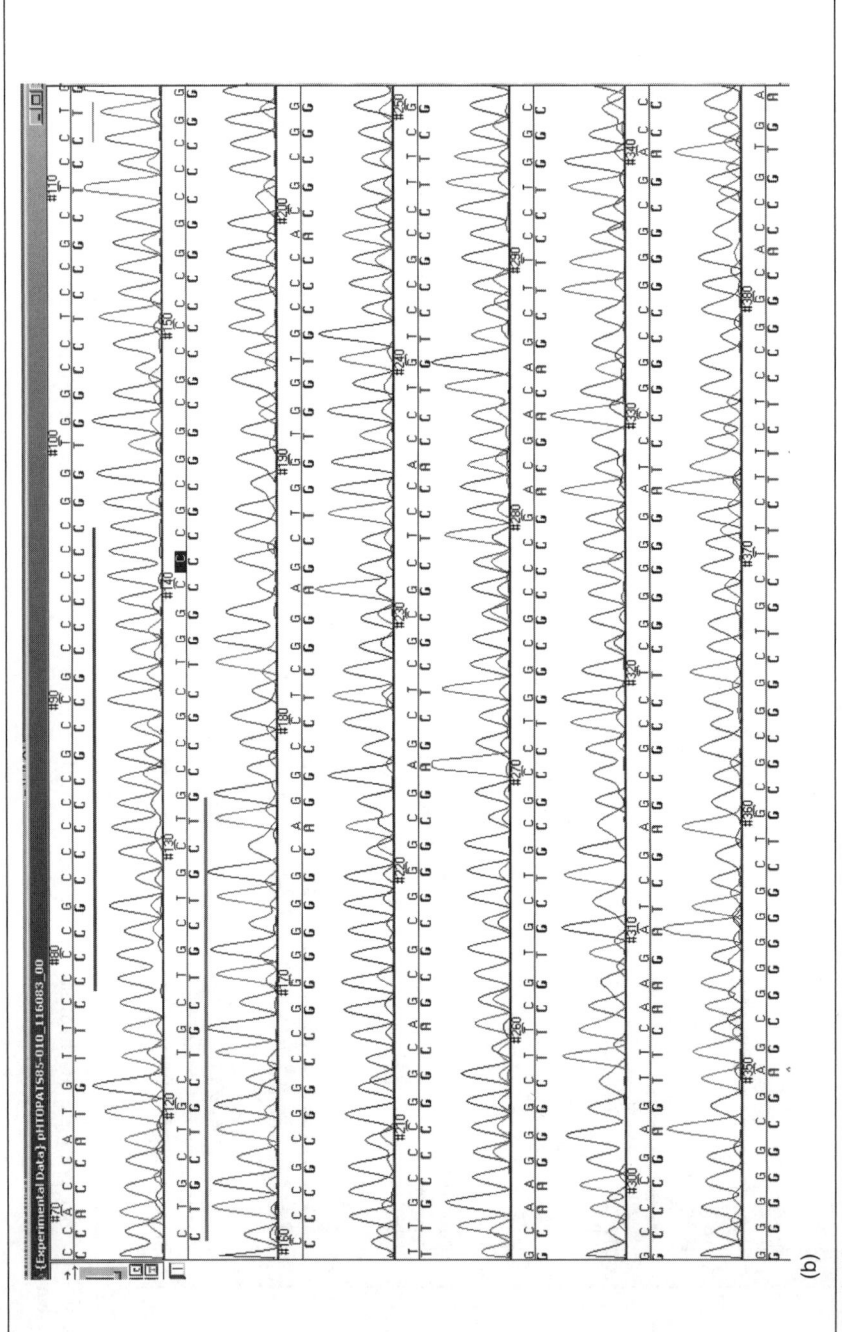

Figure 2-4(b). *Continued*

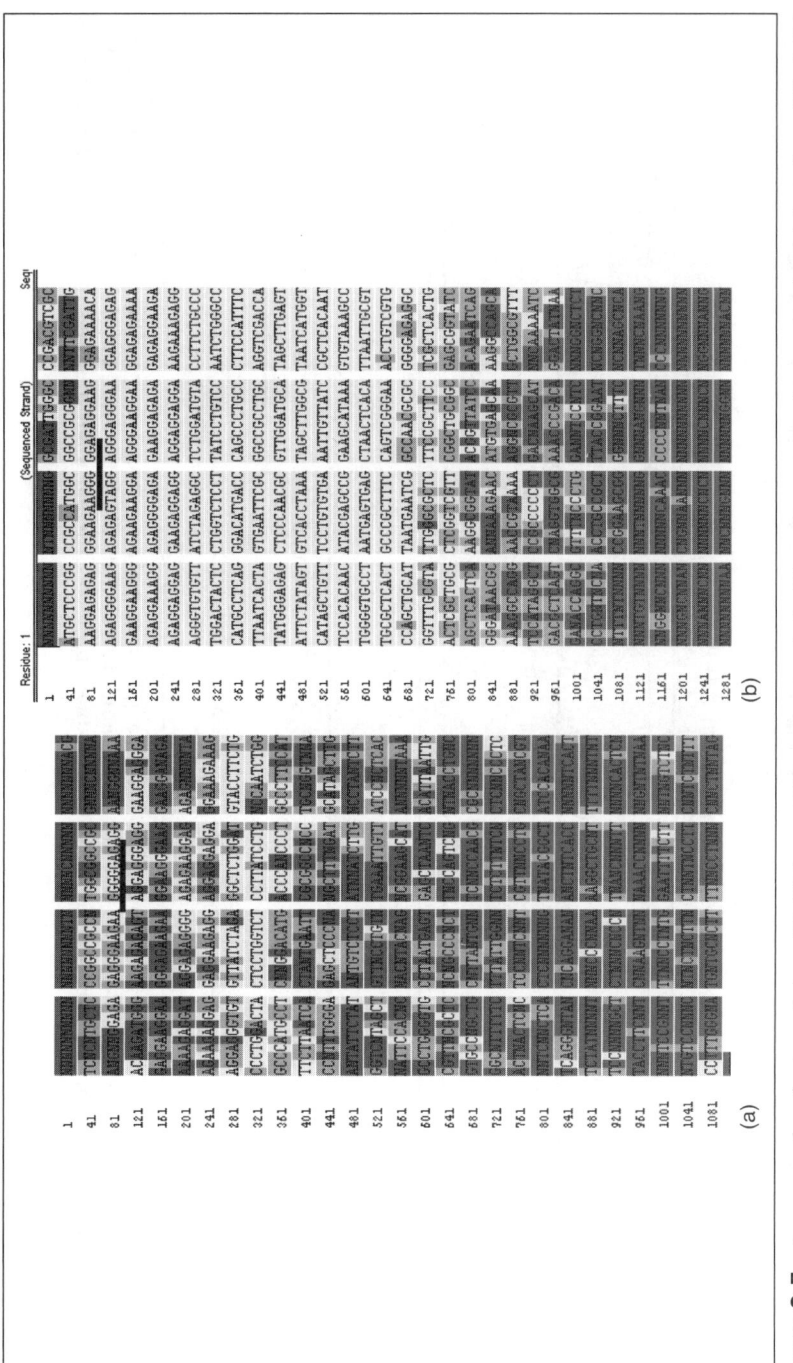

Figure 2-5. An example of a template with 59-bases GA non-repeat stretch sequenced using a standard (a) and modified (b) protocol. (see Plate 3 in the Color Addendum.) These views represent quality scores as displayed by Sequencher program. The light gray color indicates bases with higher Q quality and darker gray color indicates bases with lower Q values. The black bar shows the region with the same sequence.

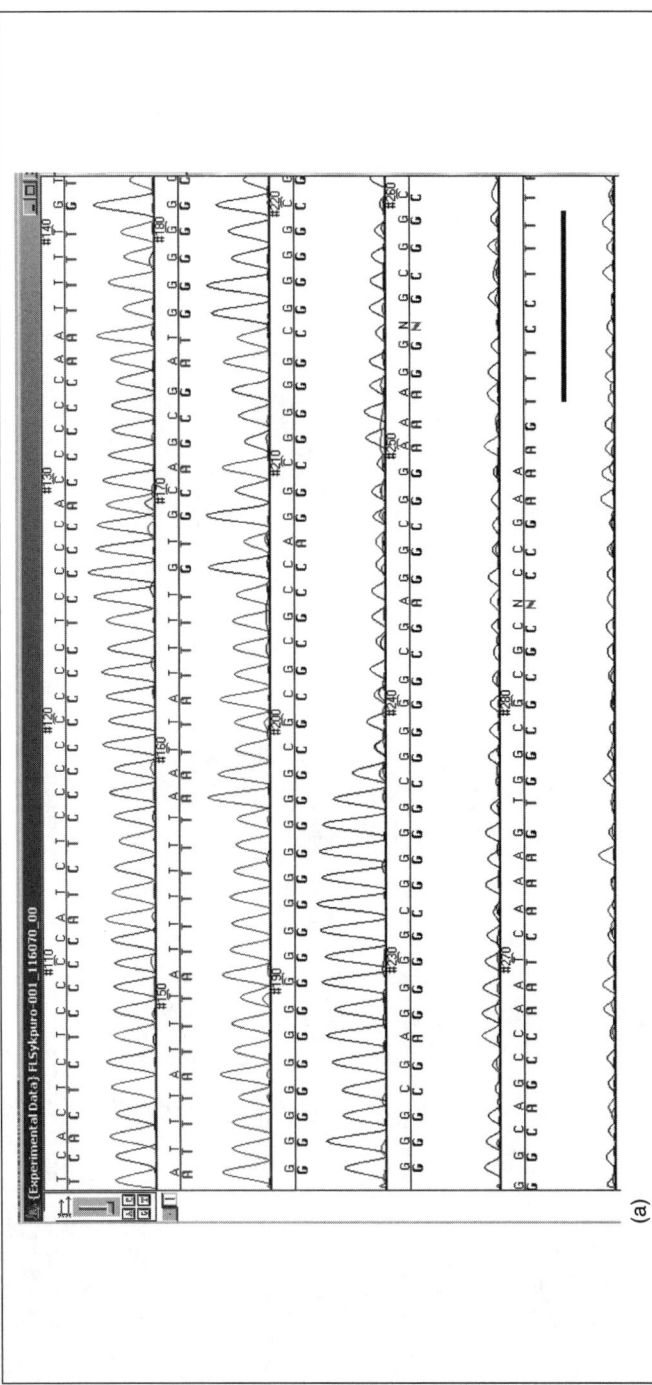

Figure 2-6. An example of sequencing through a 37-base-long G homopolymer stretch using either standard (a) or modified (b) protocol. Notice that the basecaller stops to call bases around base 280 (standard protocol) but the sequence is still readable even after base 400 for a chromatogram obtained with a modified protocol. The black bar indicates the same sequence in both traces.

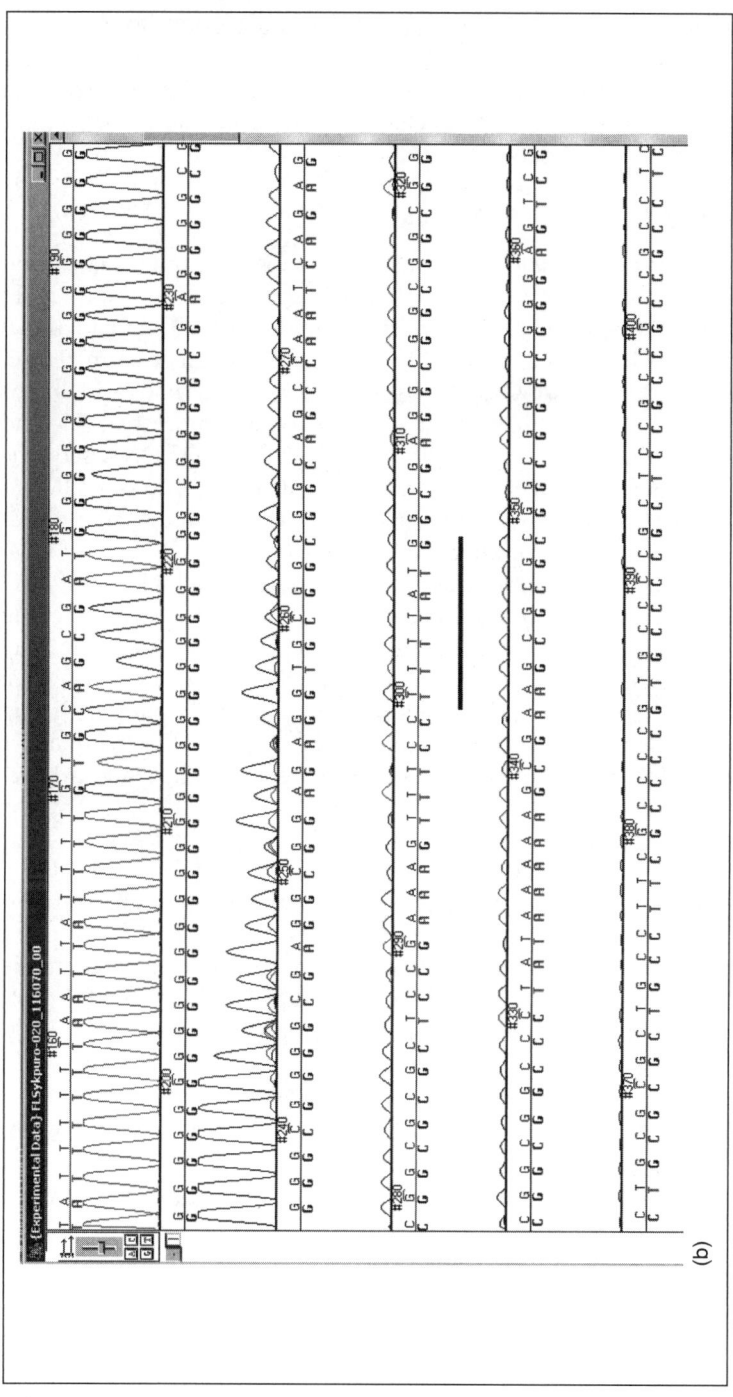

Figure 2-6(b). *Continued*

Examples of Sequencing of Various Difficult Templates

In this section, we describe possible modifications of DNA sequencing protocols that present the best chance(s) to sequence through specific types of difficult templates.

Sequencing of GC-Rich Templates

This category of templates is arguably the most commonly encountered in a typical sequencing laboratory (DSRG survey 2005 and data in Table 2-1). Most often, the solution of choice is to add DMSO to the final concentration of 2.5% to 5%, and it seems to be effective in templates with up to 60% to 72% GC-content (2–4). On the other hand, the inclusion of a five-minute heat denaturation step is even more effective (15) and, of course, does not require the addition of DMSO. In our DNA sequencing core laboratory, we have developed protocols (built into our LIMS) that automatically switch between various chemistries depending on the GC content of a particular region in a template. When the GC content is ≤70%, just heat denaturation alone is sufficient to sequence a majority of difficult regions. When the GC content is between 71% and 80%, we add reagent A (from a panel of seven Rx reagents sold by Invitrogen, Inc.) or 1 M betaine (Sigma-Aldrich, St. Louis, MO). For templates with a GC content from 81% to 90%, we use dGTP V3.0 chemistry in conjunction with reagent A or C. It is worth remembering that dGTP Big Dye™ occasionally produces band compression and needs to be corrected using standard BigDye chemistry (16). Above 90% GC content, we (through our LIMS) apply the two-step process that includes PCR amplification of the fragment in question, using 7-deaza-dGTP instead of dGTP, followed by purification and sequencing using standard dye-terminator chemistry. A similar two-step approach—the only successful one thus far—also is used for sequencing through the 28-base pair hairpin structure in the ITR sequence in adeno-associated viruses. Recently we have started to evaluate the SFK protocol from GE Healthcare (26) to sequence through some very difficult templates; this appears to be a very useful tool, especially for very GC-rich templates (see also Chapter 4). A few attempts to use Tween-20 and NP-40 detergents to help with sequencing of these types of templates were unsuccessful.

Table 2-1 contains examples of sequencing through several GC-rich templates.

Sequencing Regions Containing Various Repeats

The most effective way to deal with many different kinds of repeats is to apply a heat denaturation step. More specifically, for AG/GA repeats, heat denaturation and reagent A are a very effective combination. For GCCCCT/TCCT and similar repeats, we used GTP chemistry with reagent

A. We generally recommend, in addition to the heat denaturation step, reagent A or betaine for repeats that contain Gs and reagent C for repeats that contain Cs. Table 2-1 has examples of sequencing through some repeat structures. Figure 2-3 (a and b) shows an example of the effect of heat denaturation on the sequence quality for the template that has 147 G/A non-repeat. Figure 2.5 (a and b) shows the dramatic difference in quality values for a template with 59 bases stretch of G/A.

Sequencing of Regions with Hairpin Structures

The detailed protocol varies depending on the length of the hairpin: the heat denaturation step alone is sufficient to sequence through most hairpins that are ≤15 bases long (16). On the other hand, Esposito et al. had to introduce a special blocking agent to overcome problems with sequencing fragments cloned in pDONR221 (9). For the standard 19-base pair hairpin structures that are currently commonly used in siRNA experiments (5, 8, 13, 27), we recommend the addition of reagent C (or A) to successfully read correct sequences. Only occasionally is dGTP chemistry necessary, for example, when GC pairs are concentrated at the one end of the stem (see also Chapter 5 in this volume). For longer hairpin structures, as for example in a 28-base pair ITR region, the two-step protocol (PCR in the presence of 7-deaza-GTP followed by standard sequencing) has to be used to sequence through such strong hairpin structures. Table 2-1 shows examples of sequencing through hairpin structures.

Sequencing Through Long Homopolymer Stretches

In many ways, sequencing through homopolymer stretches is the most challenging. Heat denaturation alone is sufficient to read through G or C homopolymers of 15 bases or less (though addition of reagent A or betaine improves the quality of data). However, if G or C stretches are longer than 19 bases, the two-step approach described earlier (or using SFK kit) may need to be used to achieve satisfactory sequence data. Figure 2-4 (a and b) shows examples for sequencing for C-stretch and CTG repeats and Figure 2-6 (a and b) shows examples for sequencing through 37 G homopolymer stretch.

    A variety of protocols are recommended for sequencing through poly A or T tails. Poly A/T B (V) N primers (where B is not A and V is not T; N is any base) are commonly used and are sometimes effective (31). The design of such a primer is important, as shown in Figure 2-7 (a and b). Another approach is to use a primer that spans the boundary of a unique sequence and the poly A/T tail (22), but we have found that this is not universally applicable. Our preliminary observations suggest that successful reading through long poly A/T stretches (especially >50) depends on the sequence that immediately follows such stretches: difficult regions

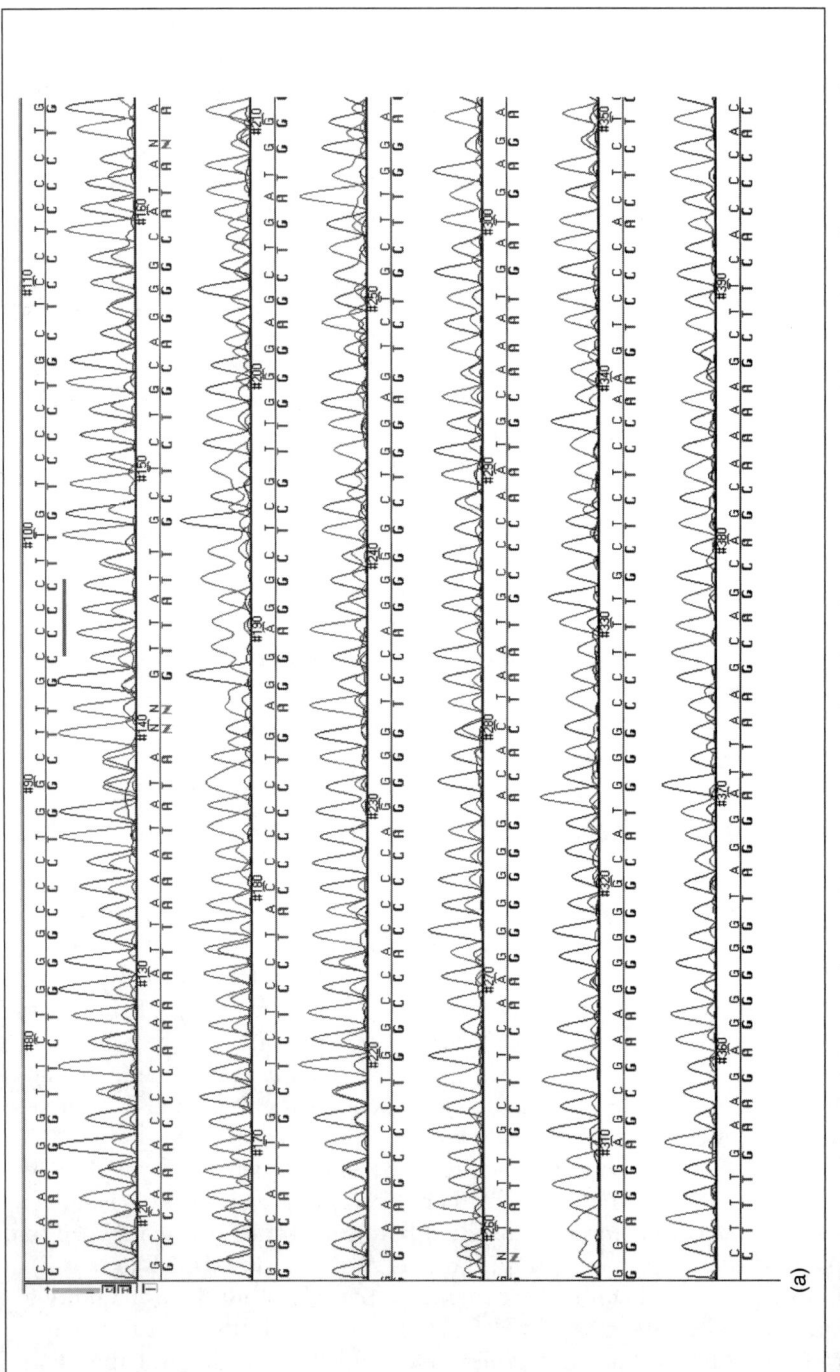

Figure 2-7. A template with long poly A tail was sequenced with either 24TVN primer (a) or 18TVN (b) primers. The gray bar indicates the common sequence.

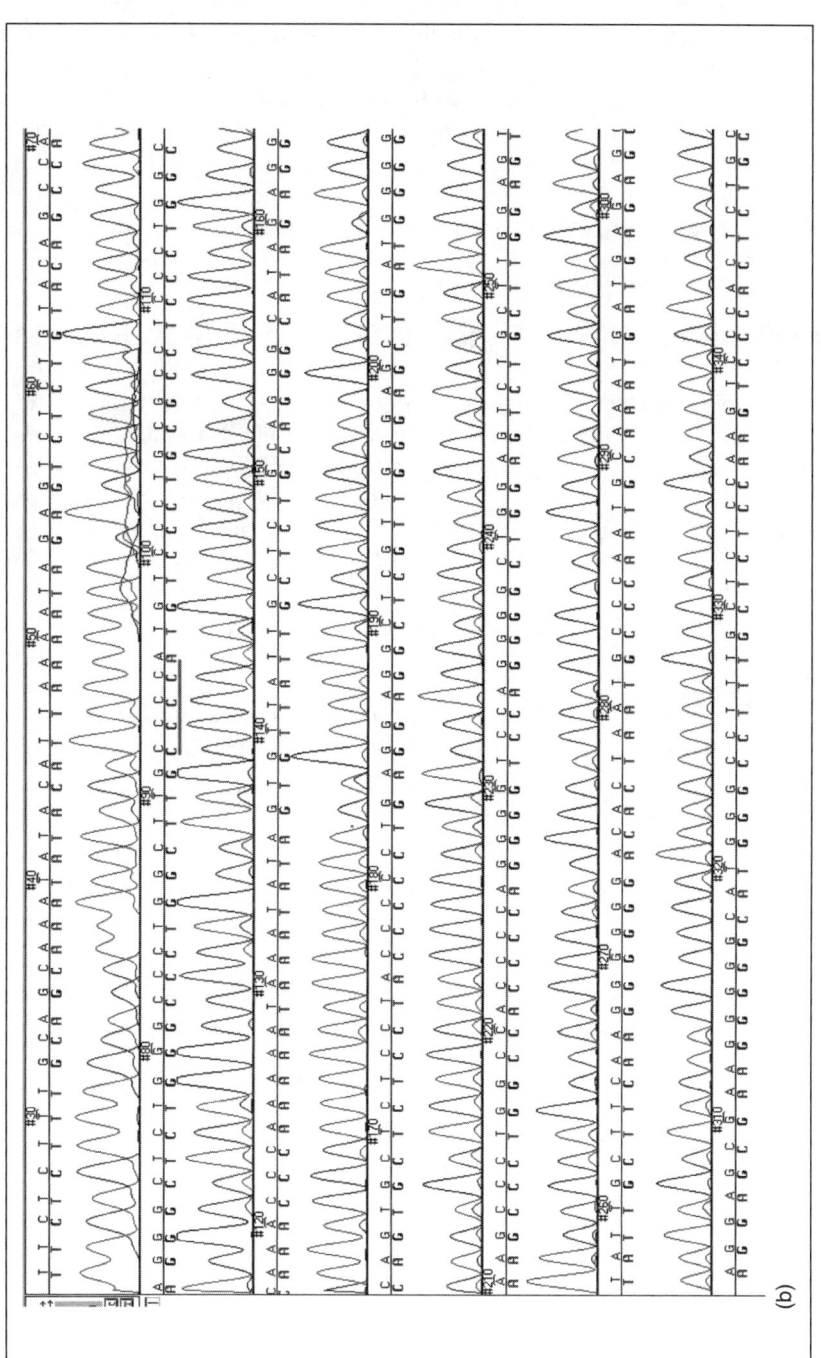

Figure 2-7(b). *Continued*

immediately following an A/T tail will make a clean sequence read harder, while it is possible to get readable A/T sequence if the downstream region is not difficult. Although not so straightforward, one way to effectively deal with homopolymer regions is to get reads from both ends, often using a new primer in reverse orientation. It follows that quite often one will not have full double-stranded coverage of such stretches. An example of such approach is shown in Figure 2-8 (see Plate 4 in the Color Addendum).

Amplification of Difficult Regions with Various DNA Polymerases

As mentioned earlier, sometimes the only effective way to sequence certain difficult regions in the DNA template is to PCR it out in the presence of 7-deaza-dGTP. Initially, we used only HiFi DNA polymerase for such amplifications. However, recently a few new DNA polymerases have become available, some even specifically targeting amplification of difficult regions (for example, Phusion and TopoTaq DNA polymerases). To assure that the optimal enzyme is used, we tested six different DNA polymerases for the ability to get through several different kinds of difficult regions. Figure 2-9 shows the example of amplification of four difficult regions using four different DNA polymerases (two other enzymes failed to produce bands for these DNAs). Based on the data presented in Figure 2-9 and in Table 2-2, we recommend using either Phusion or TopoTaq DNA polymerases when amplifying difficult regions, especially those with high GC content or containing strong hairpins. It would be worthwhile to test new Taq DNA polymerases (particularly those with high nucleotide incorporation fidelity) as they become available for their ability to produce clean and sequenceable products.

Isolation of Difficult Templates with Various Commercial Kits: Effect on Sequence Quality

It is well established that the most important factor influencing good sequencing data is the quality of the DNA preparation, assuming the amount of DNA is sufficient. As the chemistry and sample treatment vary among commercial kits, it is reasonable to expect that there will be differences in quality of sequence data obtained when the same DNA is prepared using different kits (10). In our studies (20), we prepared 12 different difficult DNAs using six conventional DNA preparation methods and a Templiphi-based protocol to evaluate if there is advantage to using one kit over another. Two well-established commercial DNA kits (from EdgeBiosystems and Qiagen) seem to be most consistent in producing high

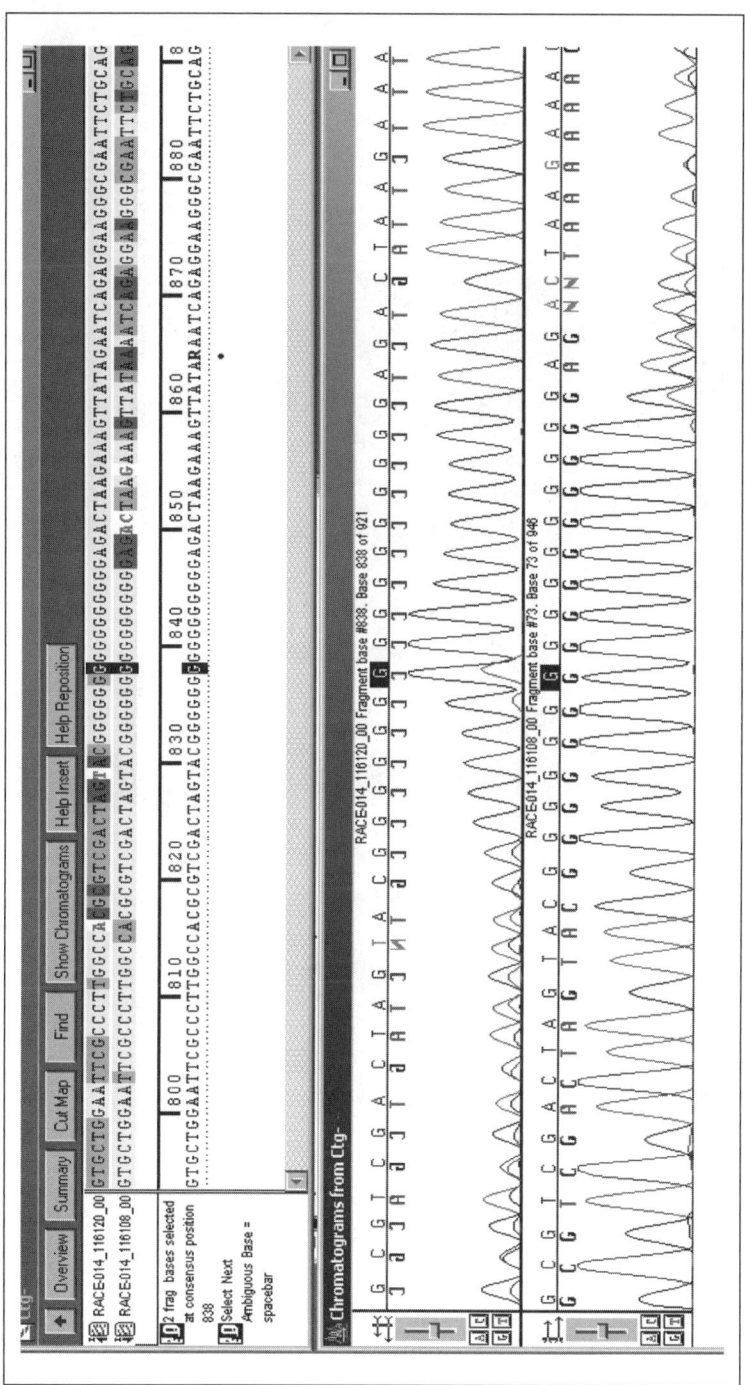

Figure 2-8. **An example of sequencing through G-stretch from both directions.** Notice that the sequence slips after 15Gs in both directions but the overall quality is satisfactory and may not need any other special treatments. A similar approach can be used for other types of homopolymer stretches.

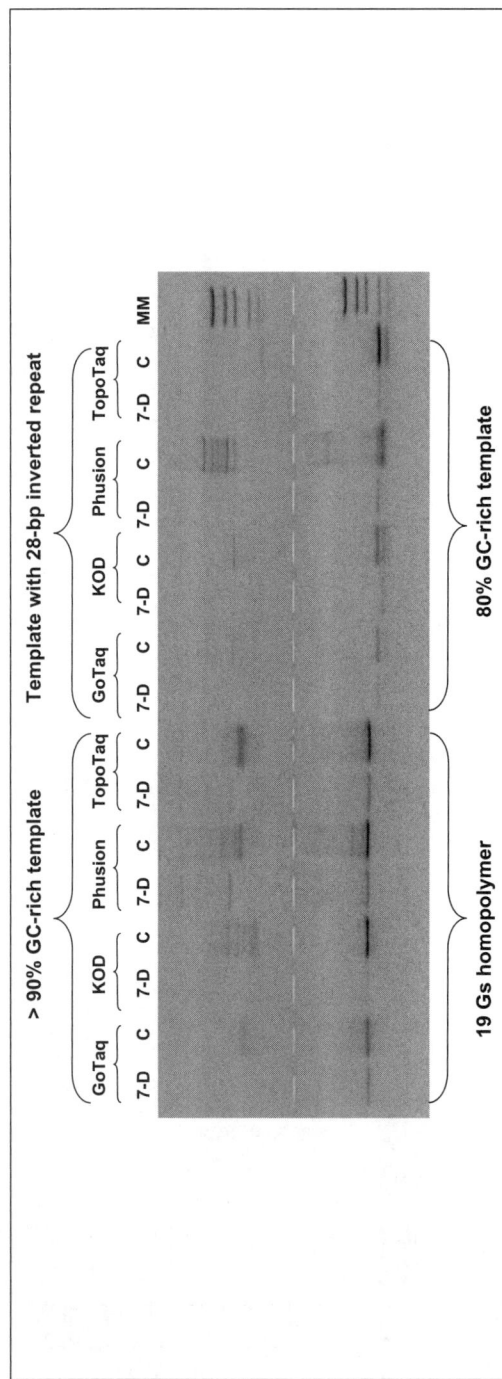

Figure 2-9. Amplification of four different difficult DNAs with four DNA polymerases. Note that the intensity of bands amplified with 7-deaza-dGTP (lanes with 7-D labels) is lighter compared to those amplified with standard nucleotides (lanes C labels). In addition, the band sizes appear different, which is due to different properties of amplicons containing either standard dGTP or 7-deaza-dGTP.

Table 2-2. **Sequence data for PCR fragments amplified using different DNA polymerases.** Each data point is the average of at least three separate reads, and read length is expressed in terms of $Q \geq 20$ values.

DNA Type	DNA Characteristics	Go Taq		KOD		Phusion		TopTaq		Vent	
		F	R	F	R	F	R	F	R	F	R
C	94% GC over 200 pb/101 inverted repeat	NB/ND	NB/ND	NB/ND	NB/ND	0	477 ± 33	161 ± 64	514 ± 11	103 ± 3	176 ± 15
7-D		NB/ND	NB/ND	NB/ND	NB/ND	457 ± 99	906 ± 40	616 ± 10	768 ± 49	721 ± 3	462 ± 13
C	28 base inverted repeat	242 ± 7	488 ± 44	447 ± 44	506 ± 13	363 ± 43	0	300 ± 39	99 ± 2	NB/ND	NB/ND
7-D		265 ± 25	471 ± 6	634 ± 94	590 ± 93	401 ± 13	525 ± 9	442 ± 64	693 ± 49	NB/ND	NB/ND
C	18 base G/C stretch and 10 G/C stretch separated by 20 bases	165 ± 6	176 ± 67	0	605 ± 20	613 ± 39	543 ± 43	615 ± 31	0	586 ± 4	0
7-D		340 ± 6	271 ± 74	0	618 ± 3	627 ± 24	651 ± 6	672 ± 3	619 ± 42	654 ± 29	0
C	30 base inverted repeat with 61 base loop and 19 G/C 24 bases away	247 ± 6	0	260 ± 2	253 ± 8	124 ± 44	139 ± 7	173 ± 8	127 ± 18	NB/ND	NB/ND
7-D		315 ± 11	0	257 ± 3	258 ± 2	218 ± 7	250 ± 8	327 ± 43	239 ± 10	NB/ND	NB/ND

DNA Polymerase

Each data point is the average of at least three separate reads. Abbreviations are: C, amplicon obtained using standard PCR conditions for any given DNA polymerase; 7-D, amplicon obtained with 7-deaza-dGTP, instead of dGTP; NB, no band visible on an agarose gel; ND, no sequence data. All reads are expressed in $Q \geq 20$ values. F/R refer to either forward or reverse primer used for sequencing.

Table 2-3. The performance of six classical plasmid preparation methods. Twelve different difficult templates were prepared using six commercial kits. Data shows how many times (No. and %) given method produced DNA that resulted in longest reads (measured with Q ≥ 20 values) using four sequencing methods. Method 1 is equivalent to a standard ABI-like protocol. Method 2 included heat denaturation only. Methods 3 and 4 were the two best methods for a given DNA template and included heat denaturation and one of additives (see reference 7 for details). The twelve templates used in this experiment are a subset of templates shown in Table 2-1.

DNA Preparation Method	DNA Sequencing Method							
	Method #1		Method #2		Method #3		Method #4	
	N	%	N	%	N	%	N	%
EdgeBiosystems	7	58.3	4	33.3	4	33.3	5	38.5
Eppendorf	0	0	0	0	0	0	0	0
Marligen	2	16.7	0	0	0	0	1	7.7
Promega	0	0	0	0	0	0	0	0
Sigma	0	0	0	0	5	41.7	4	30.8
Qiagen	3	25.0	8	67.3	3	25.0	3	23.0

N, number of times the method gave the best results.

quality DNAs leading to longest reads (20), as shown in Table 2-3. The Templiphi produces branched and somewhat unstructured molecules (6, 24) that we were expecting would be easier to read through, but unfortunately this turned out to be incorrect, as seen in Table 2-4. Lowering the amplification temperature when using Templiphi from 30° to 10°C helps, but no sufficient data yet exist (J.K., unpublished observation). The technical literature from Amersham (26) (currently part of GE Healthcare) claims that certain types of difficult templates, prepared using modified Templiphi (SFK kit), are easier to sequence compared to those isolated with conventional methods. An extensive description of the SFK (called the TempliPhi Sequence Resolver Kit) methodology is in Chapter 4.

The World Wide Web as a Resource for Information on Sequencing of Difficult Templates

In the era of Internet data search engines such as Yahoo and Google, it is only natural that one would be tempted to scan the World Wide Web for

Table 2-4. **Comparison of read lengths between DNAs prepared using either Templiphi or Qiagen/Marligen methods.** The Templiphi-derived templates were prepared as described in references 6 and 24. Method 1 is equivalent to a standard ABI-like protocol. Method 2 included heat denaturation only. Method 3 was the best protocol for a given template and included heat denaturation and one of additives (see reference 7 for details). The templates used in this experiment are a subset of templates shown in Table 2.1 Read length is expressed in terms of $Q \geq 20$ values.

DNA #	Characteristics	Preparation Method	DNA Sequencing Method		
			#1	#2	#3
1	CCT/TTTCCC	TempliPhi	340	515	594
		Qiagen	375	495	647
2	TCC/GCC	TempliPhi	635	640	626
		Qiagen	599	659	642
3	CTT/CCCT/CCTT	TempliPhi	0	0	0
		Marligen	337	458	454
4	Alu-repeats	TempliPhi	128	147	292
		Qiagen	214	240	259
5	90% GC	TempliPhi	111	278	285
		Marligen	0	292	327
6	88% GC	TempliPhi	143	165	181
		Qiagen	313	502	587
7	75% GC	TempliPhi	148	610	567
		Qiagen	565	642	600
8	63% GC	TempliPhi	524	591	596
		Qiagen	556	575	596
9	63% GC + strong hairpin	TempliPhi	<100	<100	<100
		Qiagen	<100	358	100
10	54% GC	TempliPhi	634	663	634
		Qiagen	640	671	643
11	shRNA hairpin 1	TempliPhi	497	487	501
		Qiagen	609	663	653
12	shRNA hairpin 2	TempliPhi	355	421	571
		Qiagen	270	343	526

information or advice regarding sequencing of difficult templates. For example, typing the key words "difficult DNA templates" into Google's search engine results in about 1,600,000 hits (411,000 in Yahoo's search engine), which lead to about 1000 sites, as of October 2007. Rarely, though, does one get any advice beyond "DMSO or glycerol" for GC-rich samples, or some unspecified proprietary treatment in a few commercial sites. Also quite unusual is the fact that almost all sites cite the same or similar amounts of DNA needed for optimal read length, concentrations of interfering agents, etc., without providing references to any published or unpublished data. An example of such unsubstantiated advice is the recommended amount of PCR fragment needed for optimal read length: use 10 ng of DNA/100 bases. In our experiments (18; see Chapter 1) we demonstrate that even for a PCR fragment of 3200 base pairs, 1 to 2 ng DNA is fully sufficient for optimal read lengths. In fact, we carried out similar experiments on PCR fragments of about 600, 900, 1500, 2100, and 3200 bases (the amount of DNA/reaction varied from 0.1 ng to over 500 ng) and the results did not support the 10 ng/100 bases claim; in all cases, 1 to 2 ng of DNA already gave the optimal read length (see Chapter 1). Following such a rule in a strict sense one would have to use about 100 ng of DNA for 1000 base-pair–long PCR fragments and, correspondingly, 300 ng for a 3000 base-pair–long fragment, which clearly is not supported by the experimental data. Interestingly, there is no similar "rule" for double-stranded plasmids, though different DNA sequencing service providers suggest increasing the amount of DNA as the size of plasmid gets larger (sometimes even to 600–800 ng for plasmids larger than 10 kbp). However, as indicated in Chapter 1 (see Figure 1-6), even for plasmids of about 17 kbp, 50 ng is entirely sufficient to obtain the optimal read length (at least under described experimental conditions).

Therefore, a reader (PC user) seeking fast Web advice concerning any aspect of DNA sequencing needs to be highly skeptical and selective, as the advice one gets may not necessarily be helpful (although in most cases it will not hurt either).

Summary

In this chapter, we reviewed historical and current methods to sequence through many types of difficult templates. Although a number of solutions were suggested over the last few years, they all seem to be quite specific to a particular type of difficult template. The possible exception is the heat denaturation modification, which appears to be more broadly applicable for several different difficult templates, and which in combination with several additives gives the best chance to obtain good quality data. Data presented in Table 2-1 show that for 18% of difficult templates adding a heat denaturation step was the only possible way to obtain any sequencing data. For 32% of templates, the improvement was on the order of 1% to 10% in read length. An increase of 11% to 50% in read length was evident for another 26% of templates, and for remaining 24% the increase in read length was on the order of 51% to 600%. It is apparent that more experiments need to be carried out to extract more general rules, and it is entirely possible that, in the interest of time, any given very difficult template may need to be sequenced using a few different chemistries in parallel.

One of the potential venues to speed the development of general rules for sequencing of difficult templates is to organize a bank of well characterized DNAs and involve the broad community in applying a range of technologies. The model of such community effort is well developed, for instance when the DNA Sequencing Research Group (DSRG) conducted a study on the effect of DMSO on sequencing of some GC-rich templates (3, 4). In collaboration with bioinformaticians, who now have extensive computational tools, it should be possible to look for more detailed correlation between sequence patterns, especially right before and after difficult regions, and the success of a given chemistry. It is also conceivable that one of the big sequencing centers, National Institutes of Health, or some other entity will establish a special unit that will investigate in greater detail the sequencing of all types of difficult templates and the "art" becomes "science."

As we approach an era of $1000-per-genome sequencing, one still-unanswered question is how these new technologies will deal with more complex regions. It is possible that difficult regions will not present any challenges as most of these methods rely on assembly of very short overlapping fragments (e.g., Illumina's 1G Genetic Analyzer, Applied Biosystems's SOLiD™ instrument, or Helicos Biosystems's HeliScope™ system) or passing a stretched single-stranded DNA through a nanopore (this technology is still in an early stage and there is no journal reference with solid data). On the other hand, 454 technology acknowledges that it has problems with obtaining reliable data for homopolymers of ≥ 8 bases (www.454.com). Hopefully, researchers who are directly involved in the development of new technologies will be able to answer such questions relatively soon.

Acknowledgments

I wish to thank Dr. Laird Bloom of Wyeth Research for his critical review of this manuscript and many valuable suggestions. The support and encouragement of the management of the Biological Technologies department of Wyeth Research is also greatly appreciated. I would like to thank the editor of *Journal of Biological Technologies* for permission to use a significant portion of a paper published in Vol. 17: 207–217, 2006.

References

1. ABI PRISM® BigDye™ Terminator v3.1 Cycle Sequencing Kit. Protocol. 2002. Part number 4337035 Rev. A. Foster City, CA: Applied Biosystems.
2. Adams, P.S., Dolejsi, M.K., Hardin, S., et al. 1996. DNA sequencing of a moderately difficult template: Evaluation of the results from a *Thermus thermophilus* unknown test sample. *BioTechniques* 21:678.
3. Adams, P.S., Dolejsi, M.K., Hardin, S., et al. 1997. Effects of DMSO, thermocycling and editing on a template with 72% GC rich area: results from the 2nd Annual ABRF sequencing survey demonstrate that editing is the major factor in improving sequencing accuracy. Ninth International Genome Sequencing and Analysis Conference. *Microb Comp Genomics* 2: 198 (abstract).
4. Adams, P.S., Dolejsi, M.K., Grills, G., et al. 1999. An analysis of techniques used to improve the accuracy of automated DNA sequencing of a GC-rich template: results from the 2nd Annual ABRF DNA Sequence Research Group study. Available at: http://www.abrf.org; search for 2nd ABRF DNA Sequence Research Group Study, 1999. Accessed October 24, 2007.
5. Brummelkamp, T.R., Bernards, R., and Agami, R. 2002. A system for stable expression of short interfering RNAs in mammalian cells. *Science* 296: 550–553.
6. Dean, R.B., Nelson, J.R., Giesler, T.L., and Lasken, R.S. 2001. Rapid amplification of plasmid and phage DNA using phi29 DNA polymerase and multiply-primed rolling circle amplification. *Genome Res* 11: 1095–1099.
7. Ducat, D.C., Herrera, F.J., and Triezenberg, S.J. 2003. Overcoming obstacles in DNA sequencing of expression plasmids for short interfering RNAs. *BioTechniques* 34: 1140–1144.
8. Elbashir, S.M., Harborth, J., Weber, K., and Tuschl, T. 2002. Analysis of gene function in somatic mammalian cells using small interfering RNAs. *Methods* 26: 199–213.
9. Esposito, D., Gillette, W., and Hartley, J.L. 2003. Blocking oligonucleotides improve sequencing through inverted repeats. *BioTechniques* 35: 914–920.
10. Flick, P.K. Plasmid preparation methods for DNA sequencing. In: Kieleczawa, J. ed. *DNA Sequencing: Optimizing the Process and Analysis.* Sudbury, MA: Jones & Bartlett; 2005: 99–115.
11. Gerstner, A., Sasvari-Szekely, M., Kalasz, H., and Guttman, A. 2000. Sequencing difficult templates using membrane-mediated loading with hot sample application. *BioTechniques* 28: 628–630.
12. Haltiner, M., Kempe, T., and Tjian, R. 1985. A novel strategy for constructing point mutations. *Nucleic Acid Res* 13: 1015–1025.
13. Hammond, S.M., Caudy, A.A., and Hannon, G.J. 2001. Posttranslational gene silencing by double-stranded RNA. *Nat Rev Genet* 2: 110–119.
14. Hattori, M., and Sasaki, Y. 1986. Dideoxy sequencing method using denatured plasmid templates. *Anal Biochem* 152: 232–238.
15. Kieleczawa, J. Sequencing of difficult DNA templates. In: Kieleczawa J. (ed). *DNA Sequencing: Optimizing the Process and Analysis.* Sudbury, MA: Jones and Bartlett Publishers; 2005: 27–34.

16. Kieleczawa, J. 2005. Simple modifications of the standard DNA sequencing protocol allow for sequencing through siRNA hairpins and other repeats. *J Biomol Tech* 16: 220–223.
17. Kieleczawa, J. Controlled heat-denaturation of DNA plasmids. In: Kieleczawa, J. (ed). *DNA Sequencing: Optimizing the Process and Analysis.* Sudbury, MA: Jones and Bartlett Publishers; 2005: 1–10.
18. Kieleczawa, J., and Bajson, K. Evaluation of methods for cleanup of DNA sequencing reactions. In: Kieleczawa, J (ed). *DNA Sequencing II: Optimizing the Preparation and Clean Up.* Sudbury, MA: Jones and Bartlett Publishers; 2006: 219–240.
19. Kieleczawa, J., and Wu, P. Prolonged storage of plasmid DNAs under different conditions: effects on plasmid integrity, spectral characteristics, and DNA sequence quality. In: Kieleczawa, J. (ed). *DNA Sequencing II: Optimizing the Preparation and Clean Up.* Sudbury, MA: Jones and Bartlett Publishers; 2006: 259–274.
20. Kieleczawa, J., Li, T., and Wu, P. Preparation of difficult DNA templates using seven different commercial methods. In: Kieleczawa, J. (ed). *DNA Sequencing II: Optimizing the Preparation and Clean Up.* Sudbury, MA: Jones and Bartlett Publishers; 2006: 1–14.
21. Kieleczawa, J. 2006. Fundamentals of sequencing of difficult templates—an overview. *J Biomol Tech* 17: 207–217.
22. Langan, J.E., Rowbottom, L., Liloglou, T., Field, J.K., and Risk, J.M. 2002. Sequencing of difficult templates containing poly (A/T) tracts: closure of sequencing gaps. *BioTechniques* 33: 276–280.
23. Maxam, A.M., and Gilbert, W. 1977. A new method of sequencing DNA. *Proc Natl Acad Sci U S A* 74: 560–564.
24. Nelson, J.R., Cai, Y.C., Giesler, T.L., et al. 2002. TempliPhi™ Phi29 DNA polymerase-based rolling circle amplification of templates for DNA sequencing. *BioTechniques* 32: S44–S47.
25. Sanger, F., Nicklen, S., and Coulson, A.R. 1977. DNA sequencing with chain-terminating inhibitors. *Proc Natl Acad Sci U S A* 74: 5463–5467.
26. Sequence Finishing Kit. 2003. Product Code 25-6401-01. Fairfield, CT: GE Healthcare.
27. Sharp, P.A. 1999. RNAi and double-strand RNA. *Genes Dev* 13: 139–141.
28. Tabor, S., and Richardson, C.C. 1987. DNA sequence analysis with a modified bacteriophage T7 DNA polymerase. *Proc Natl Acad Sci U S A* 84: 4767–4771.
29. Tabor, S., and Richardson, C.C. 1990. DNA sequence analysis with a modified bacteriophage T7 DNA polymerase: effect of pyrophosphorolysis and metal ions. *J Biol Chem* 265: 8322–8328.
30. Tabor, S., and Richardson, C.C. 1995. A single residue in DNA polymerases of the *Escherichia coli* DNA polymerases I family is critical for distinguishing between deoxy and dideoxyribonucleotides. *Proc Natl Acad Sci U S A* 92: 6339–6343.
31. Thomas, M.G., Hesse, S.A., McKie, A.T., and Farzaneh, F. 1993. Sequencing of cDNA using anchored oligo dT primers. *Nucleic Acid Res* 21: 3915–3916.

32. Toneguzzo, F., Glynn, S., Levi, E., Mjolsness, S., and Hayday, A. 1988. Use of chemically modified T7 DNA polymerase for manual and automated sequencing of supercoiled DNA. *BioTechniques* 6: 460–469.
33. Yamakawa, H., Nakajima, D., and Ohara, O. 1996. Identification of sequence motifs causing band compressions on human cDNA sequencing. *DNA Res* 3: 81–86.
34. Yie, Y., Wei, Z., and Tien, P. 1993. A simplified and reliable protocol for plasmid DNA sequencing: fast miniprep and denaturation. *Nucleic Acid Res* 21: 361.
35. Zhao, X., Haqqi, T., and Yadav, S.P. 2000. Sequencing telomeric DNA templates with short tandem repeats using dye terminator cycle sequencing. *J Biomol Tech* 11: 111–121.

Color Addendum

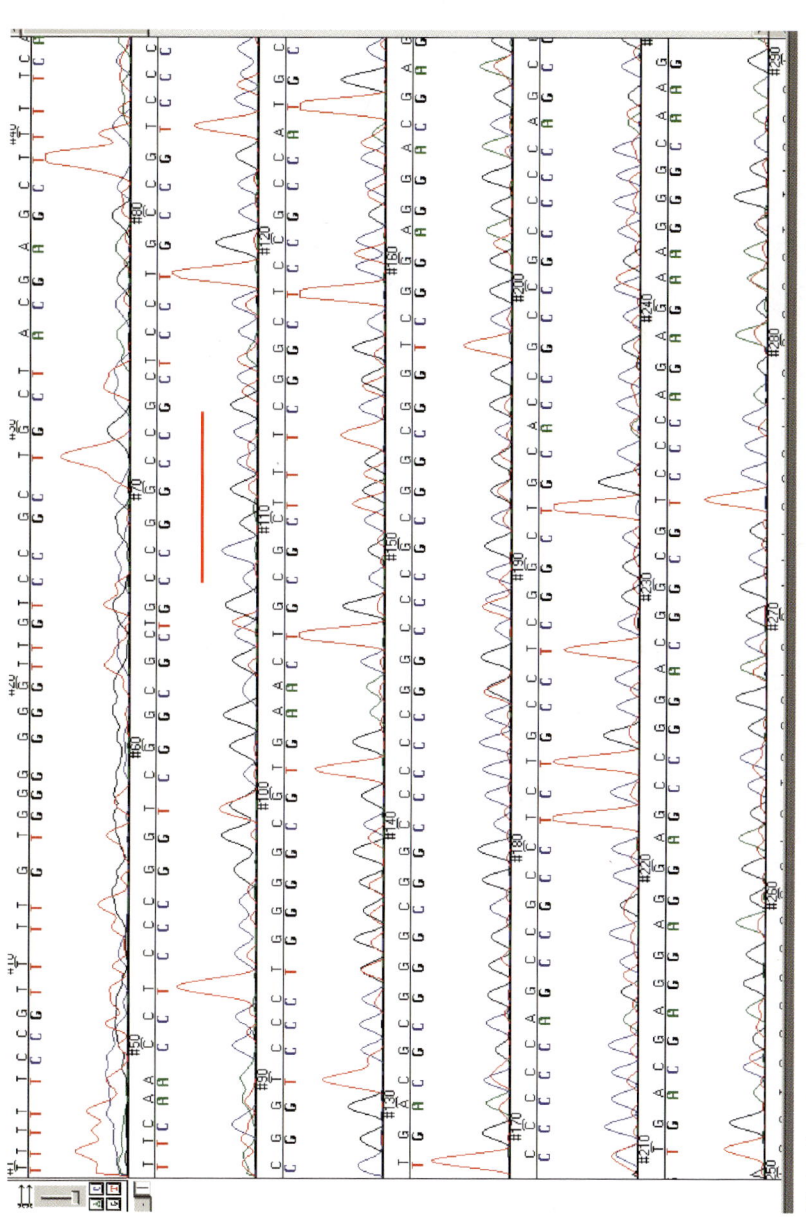

Plate 1. (Figure 1-16a; page 18)

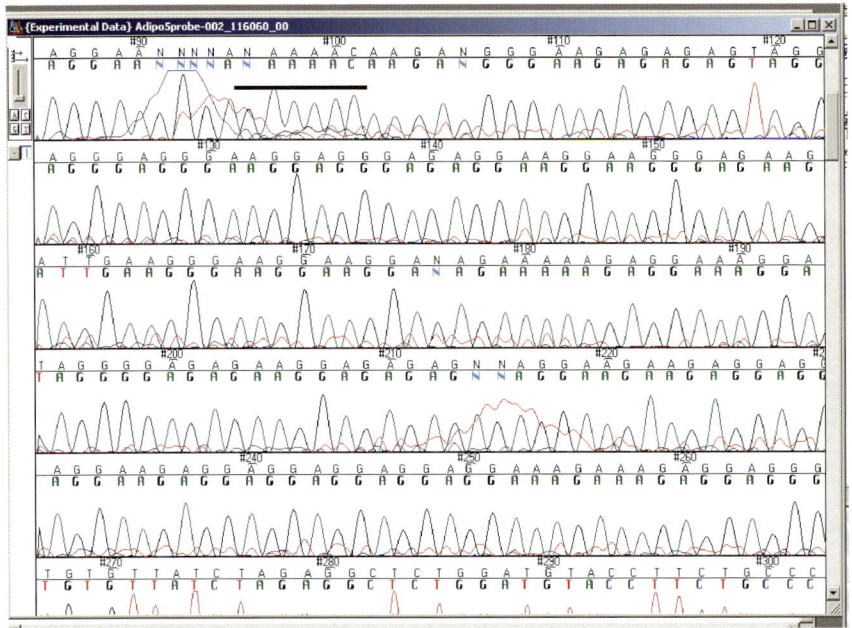

Plate 2. (Figure 2-3a; page 43)

Plate 3. (Figure 2-5a and b; page 47)

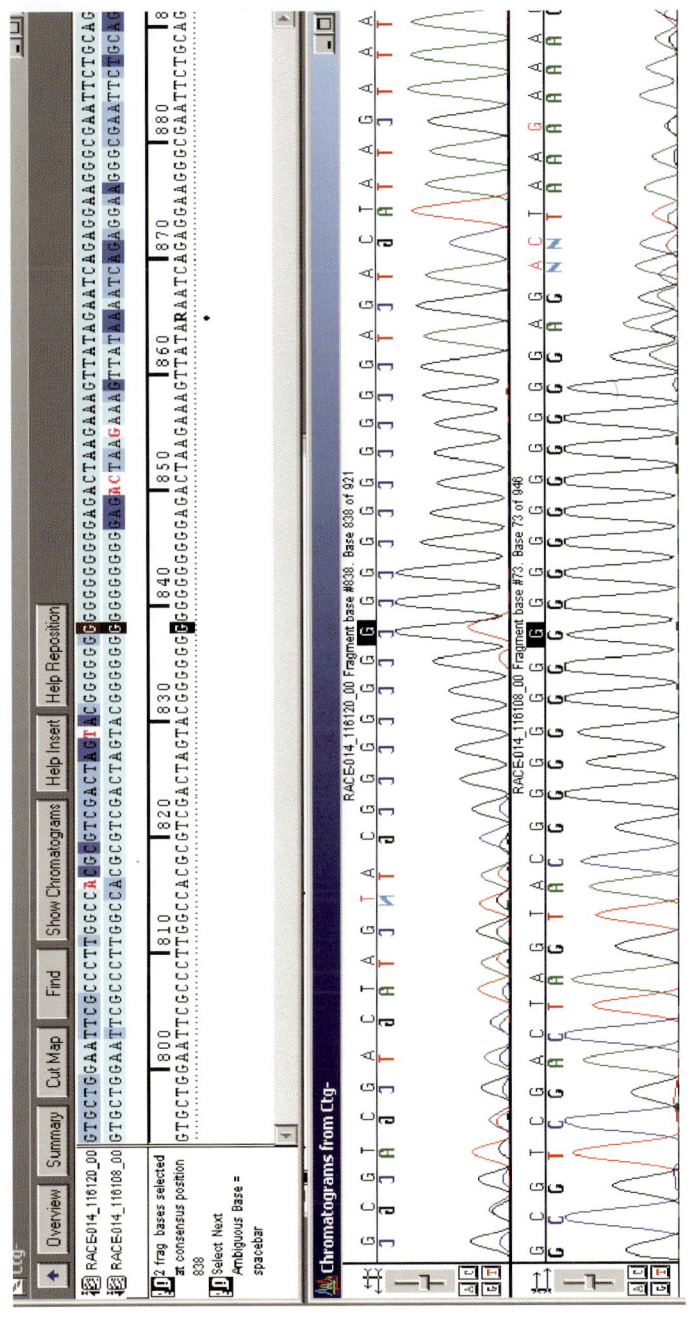

Plate 4. (Figure 2-8; page 55)

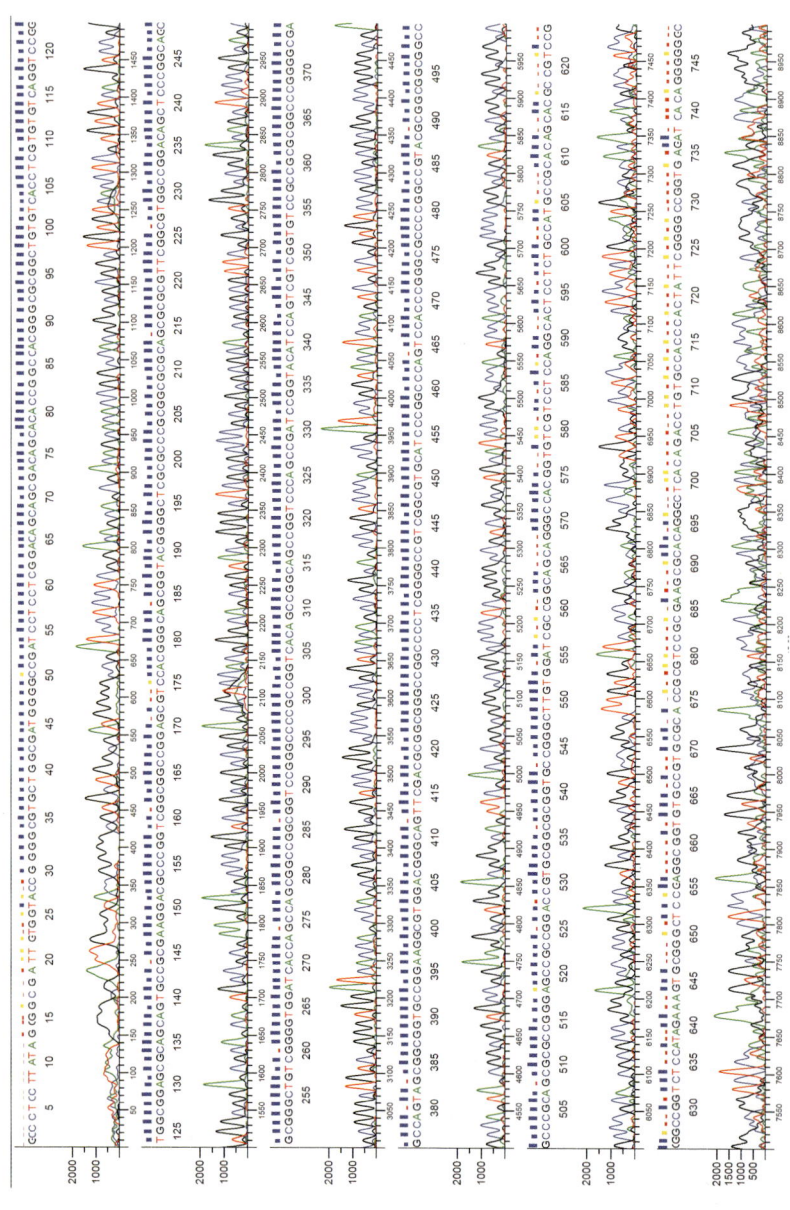

Plate 5. (Figure 3-1; page 67)

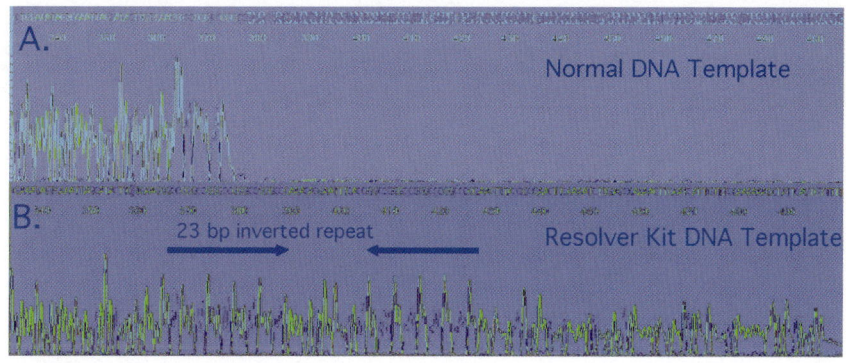

Plate 6. (Figure 4-1; page 96)

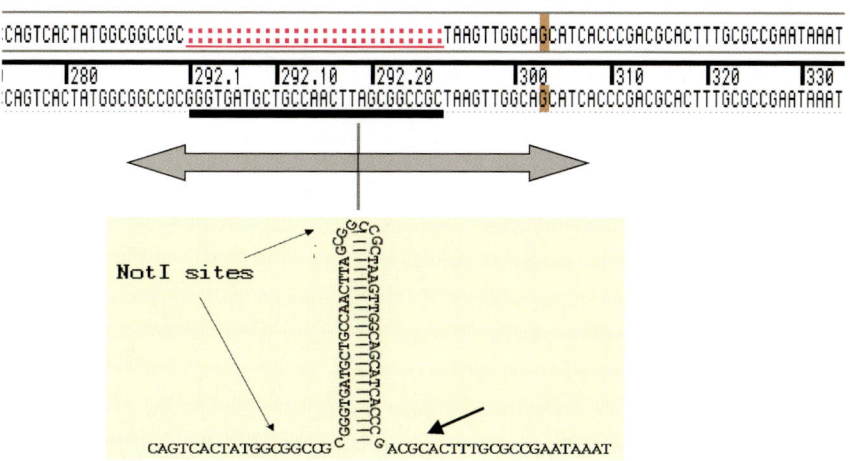

Plate 7. (Figure 5-1; page 111)

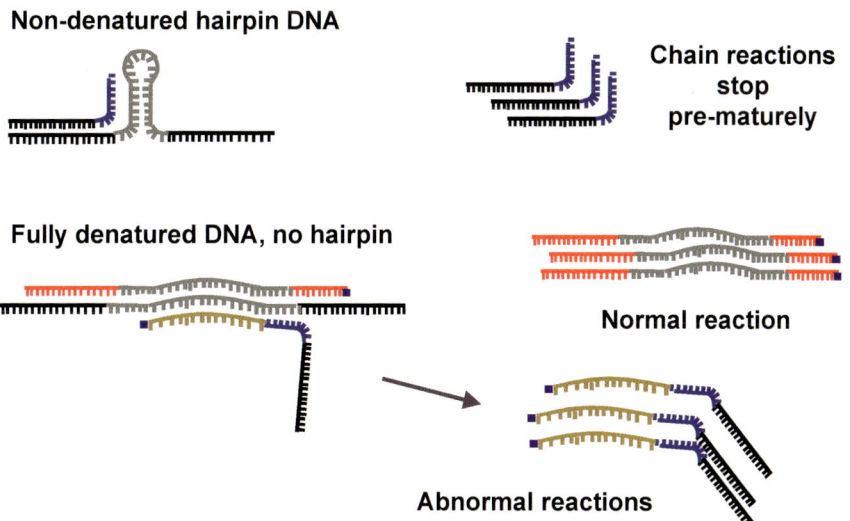

Plate 8. (Figure 5-5; page 115)

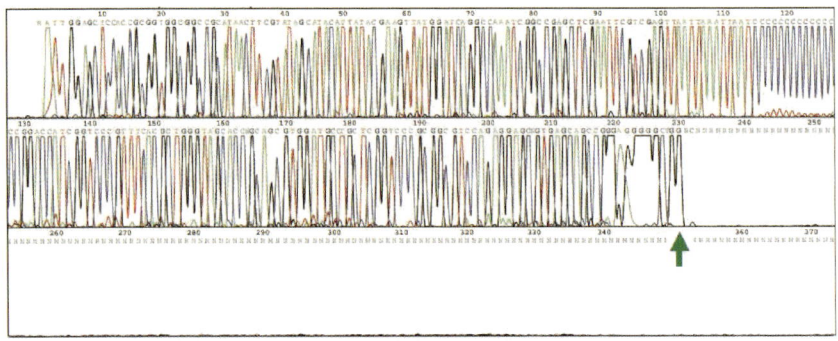

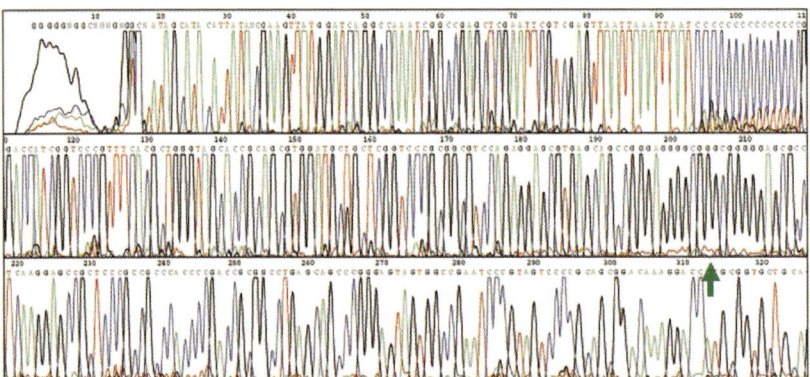

Plate 9. (Figure 6-3; page 133)

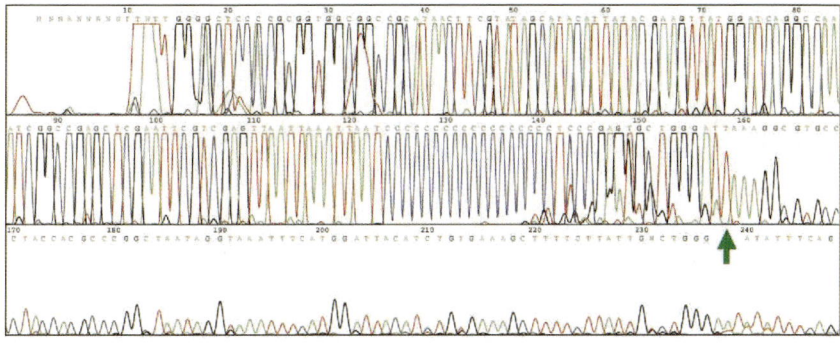

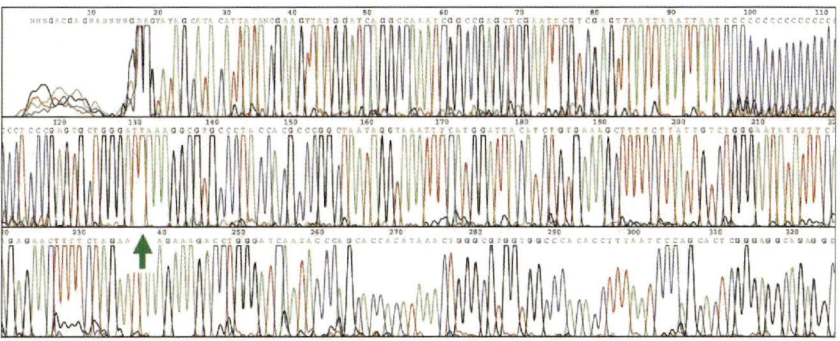

Plate 10. (Figure 6-4; page 134)

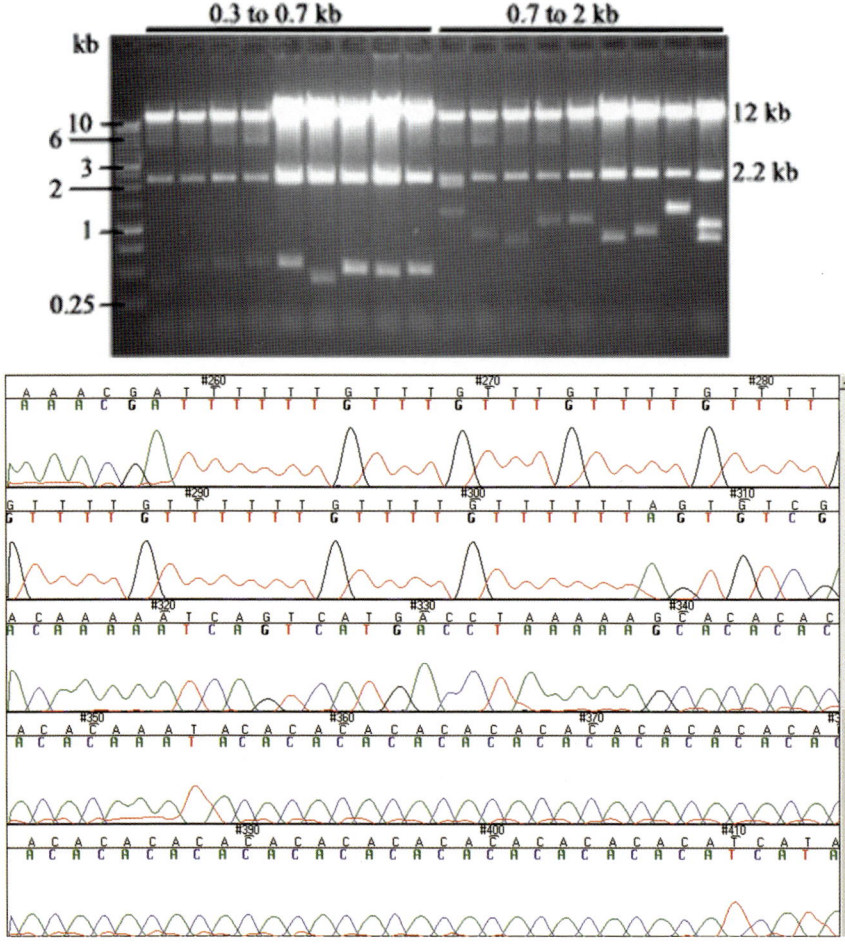

Plate 11. (Figure 7-4; page 150)

3 | Solutions for Sequencing Difficult Regions

Alicia Yang

Applied Biosystems, Foster City, CA

Despite significant advances in both sequencing chemistry and automated sequencing technology, certain base patterns remain difficult to sequence. Researchers often must seek solutions beyond standard protocols when they encounter DNA regions that are GC rich, have various polynucleo-tide-repeats, or contain extended homopolymers. Each of these DNA regions poses a different problem during sequencing. For example, GC-rich regions are hard to sequence because of their higher melting temperature requirement (3, 6–10). Sequences that contain long (>20) homopolymers sometimes result in enzyme slippage that causes a stuttering effect in fluorescent sequencing reactions. One hopes to find a solution that would resolve all these difficult-to-sequence contexts. However, the varying nature of the challenges requires varying solutions.

Applied Biosystems has been the leader in sequencing chemistry for almost two decades. Recently, we have endeavored to develop a finishing kit that will resolve most difficult sequences. Rather than concentrating on a single sequencing chemistry for all templates, we identified targeted strategies for successfully resolving the various types of difficult-to-sequence templates, along with some new formulations for those especially tough regions. In our study, we focused on what we believe, based on customer feedback, are the most common problematic regions: GC-rich regions, polynucleotide repeats, and homopolymer regions.

GC-Rich Regions

GC-rich regions usually refer to templates in which more than 65% of the bases are guanosine (G) and cytosine (C) bases (10). These templates have

DNA III: Dealing with Difficult Templates
Edited by Jan Kieleczawa
©2008 Jones and Bartlett Publishers

higher melting points because of G's and C's tendency to form triple hydrogen bonds with their complementary strands. GC-rich templates also can form these triple bonds with GC bases within the same strand, leading to stable secondary structures. Difficulty in sequencing these templates comes from the stability of these triple bond formations (6). Researchers hypothesize that stable secondary structures can "prohibit the denaturation, annealing and extension step during PCR process, subsequently resulting in inefficient DNA sequencing" (5, 8). The same concept can be applied to plasmids during cycle sequencing: stable secondary structures formed by these GC-rich regions can prohibit complete denaturation, limiting polymerase access to template binding, resulting in short sequence or low signal products (Figure 3-1) (see Plate 5 in the Color Addendum).

Polynucleotide Repeats

Polynucleotide repeats refer to regions containing 2, 3, 4, or a higher number of base repeat elements, such as AT, GT, CT, CAA or other repeats. Repeats containing adenine (A) or thymine (T) bases usually sequence through. Repeats containing G/C bases often pose problems. One possible explanation is the aforementioned triple hydrogen bond formation of the G and C bases, which hinders strand denaturation and polymerase binding and processing. The resulting sequence problems range from an abrupt stop in sequence to a gradual decrease in signal (Figure 3-2).

Homopolymer Regions

Homopolymer regions, as the name implies, describe a region of DNA that contains a repeating string of a single nucleotide. This homopolymer region can range in length from as few as 5 bases to as many as 70 bases. The major difficulty with sequencing through homopolymer regions lies in polymerase's processivity. Each time the enzyme falls off the enzyme-template duplex, the molecules are allowed to "breathe" and may not be in correct register when the enzyme re-binds and continues to extend (Diane Bond, personal communication). This "slipped" extension results in secondary sequences that appear below the primary sequence after the homopolymer region. The secondary sequence often resembles the primary sequence except it is one or two bases shifted left or right. The enzyme slippage problem in homopolymer regions is even more pronounced in PCR products. Because PCR amplifies templates in a logarithmic manner, each newly created "slipped" template gets copied again, thus amplifying the slippage problem. The end result, when sequenced, is a series of overlapping fan-shaped fluorescent sequences (Figure 3-3).

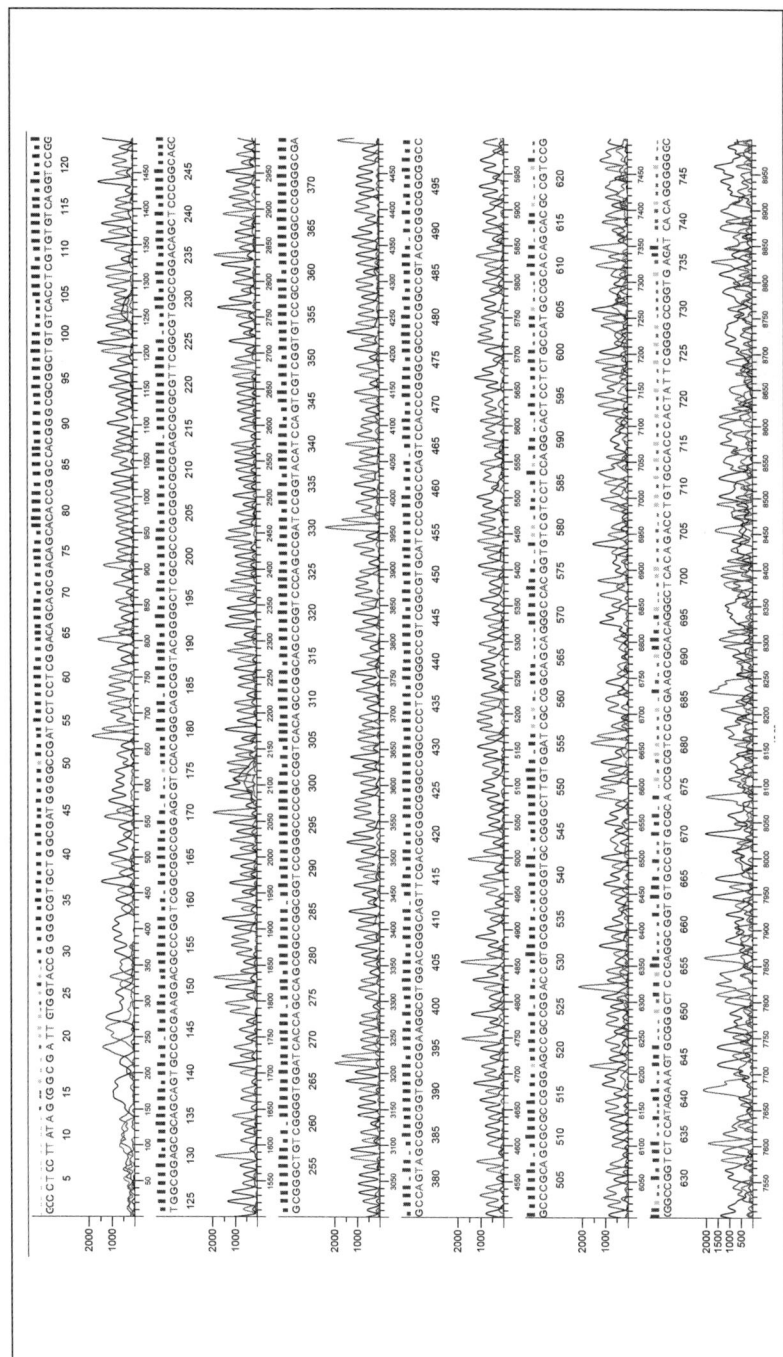

Figure 3-1. Sequence of GC-rich (74%) template showing low signal and noisy baseline, leading to shortened quality basecalls (see Plate 5 in the Color Addendum). The average raw signal intensity for each base are in the low hundreds (A:96, C:123, G:182, T:107), which is significantly below the acceptable signal levels of at least 1000. Numbers above each peak indicate base numbers. Numbers below each peak indicate the scan numbers.

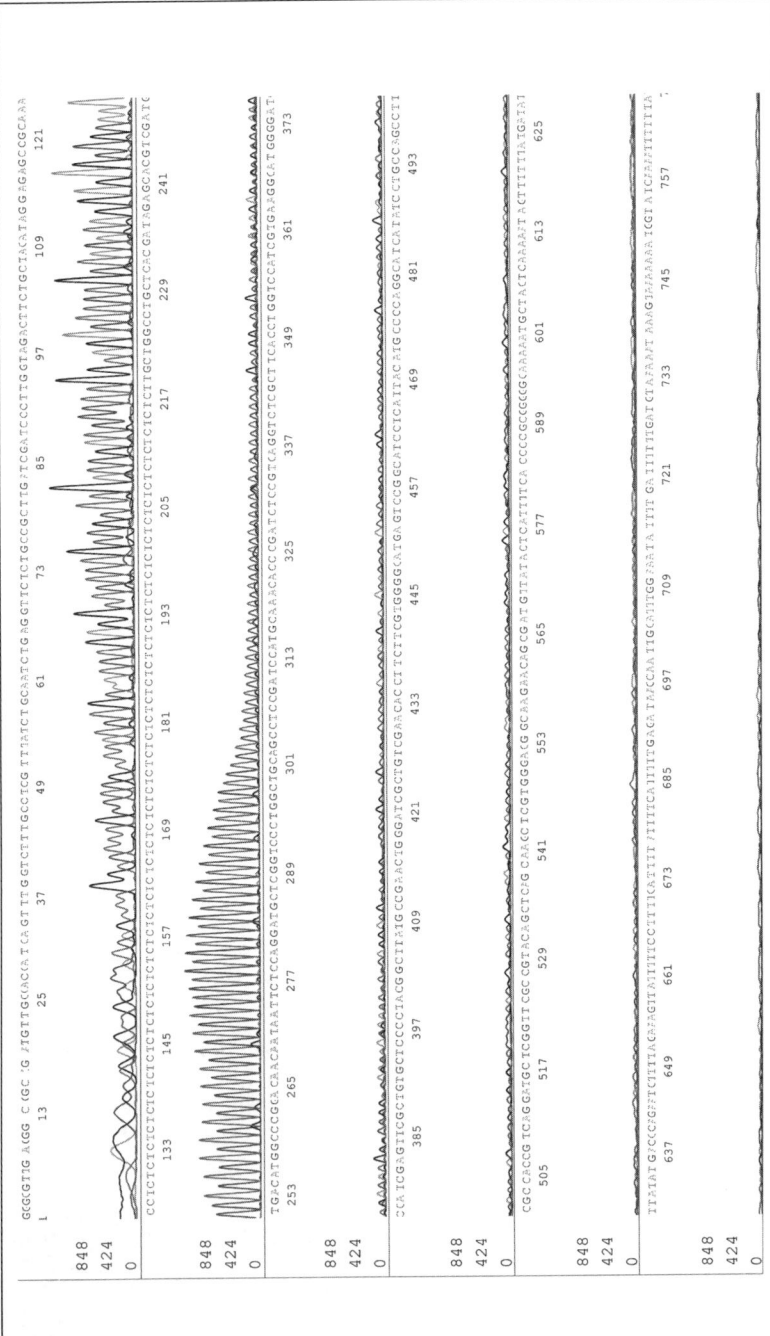

Figure 3-2. Sequence of dinucleotide repeat template showing early termination. Although software tries to basecall peaks after the significant signal drop at around base 180, signals for individual bases after this point are so low that confidence of correct basecall is also low.

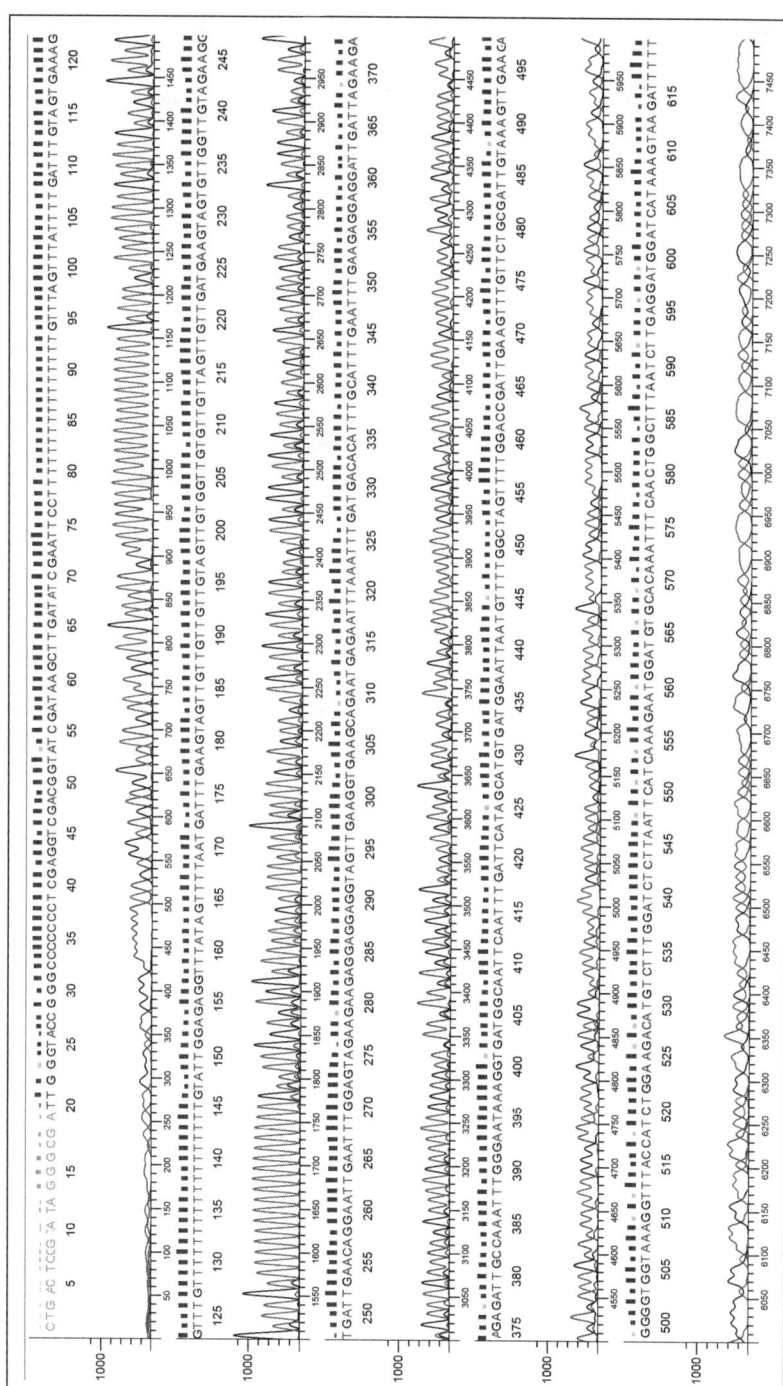

Figure 3-3. Sequence of template with homopolymer region showing enzyme slippage. N+1 peaks can be seen after the first homopolymer region, although only slightly. However, these secondary peaks start to show more prominently after the second homopolymer region.

Solutions

Over the years Applied Biosystems has developed fluorescence-based sequencing kits to address many issues associated with sequencing of the aforementioned difficult regions, including the BigDye® Terminator v3.1 Cycle Sequencing Kit (for GC-rich regions), the dGTP BigDye® Terminator v3.0 or v1.0 Cycle Sequencing Kit (for GT- and G-rich regions), and the ABI PRISM® dRhodamine Terminator Cycle Sequencing Kit (for homopolymers). The approach we took for this study was to use existing Applied Biosystems Cycle Sequencing kits, along with simple adjustments such as altering cycling conditions or primer design to improve difficult templates for sequencing success. Our goal was to provide users with simple strategies that are easy to implement and produce good results. In this chapter, we further evaluate enhancements to the existing Applied Biosystems kits that improve their ability to sequence difficult regions.

Materials and Methods

Formulations

Seven Applied Biosystems sequencing chemistries were used in this study:

1. BigDye® Terminator v3.1 Cycle Sequencing Kit (p/n 4337457, 5000 rxn kit)
2. dGTP BigDye® Terminator v3.0 Cycle Sequencing Kit (p/n 4390229, 100 rxn kit)
3. ABI PRISM® dRhodamine Terminator Cycle Sequencing Kit (p/n 403044, 100 rxn kit)
4. BigDye® Terminator v3.1 kit/dGTP BigDye® Terminator v3.0 kit, 75:25 Blend (1)
5. Experimental Formulation F1
6. Experimental Formulation F2
7. Experimental Formulation F3

BigDye® Terminator v3.1 Sequencing Kit is Applied Biosystems' (AB) flagship sequencing chemistry kit. This chemistry is the most robust and versatile kit out of the suite of sequencing chemistry kits offered by AB. More specialized chemistry kits, such as dGTP BigDye® Terminator v3.0 kit and ABI PRISM® dRhodamine kit, complete the AB sequencing chemistry product line. Experimental formulations are company confidential and thus cannot be discussed in detail.

Blending

The blending method tested here is the method proposed by several contributors to ABRF where a blend of BigDye® Terminator v3.1 Ready Reaction Mix and dGTP BigDye® Terminator v3.0 Ready Reaction mix in a ratio of 75:25 is used in the sequencing reaction (G. Grills, unpublished observation; 7, 10). The two-kit blend is made first into a master mix and used in the same manner as regular Ready Reaction Mix in sequencing reactions. A normal 1× sequencing reaction using this blend would require 8 µL of the 75:25 blended master mix in a 20 µL sequencing reaction volume.

Dilutions

All seven formulations were tested in dilution series. Dilutions are defined by the amount of BigDye Terminator Ready Reaction mix (RR) added to the reaction. A 1× reaction is defined as 8 µL RR in a 20 µL reaction volume. A series of dilutions were tested. All RR mixes are diluted in appropriate amounts of BigDye® Terminator v3.1/v1.1 5× Sequencing Buffer (2).

Templates

Experiments were conducted on 20 different templates containing difficult-to-sequence motifs. This selection represents the three most common difficult regions discussed previously: GC-rich region, polynucleotide repeats, and homopolymers. Each template chosen contains a different type and degree of difficulty in sequencing. Some may contain a single type, such as GC-rich regions, whereas others may contain homopolymers in one region and dinucleotide repeats in another. All templates were sequenced with the seven formulations described previously and at all aforementioned dilution levels.

Cycle Sequencing Conditions

For each experiment, four replicates of each configuration were processed using the following cycling condition as specified in the BigDye® Terminator v3.1 kit protocol booklet (2):

96°C	hold for 1 min	
96°C	for 10 sec	
50°C	for 5 sec	25 cycles
60°C	for 4 min	
4°C	hold	

Electrophoresis

Sequencing reactions were purified using EDTA/ethanol precipitation protocol specified in BigDye® Terminator v3.1 kit protocol booklet (2). Purified samples are then resuspended in 10 μL Hi-Di™ Formamide (Applied Biosystems) prior to loading onto an Applied Biosystems 3730*xl* Genetic Analyzer capillary electrophoresis instrument (1). Capillary electrophoresis was carried out according to standard instructions that accompany the *User Guide for the Applied Biosystems 3730xl Genetic Analyzer* (1).

Analysis Methods

Data for this study were analyzed with six different metrics as defined and described by Bond et al. (3). These are automated tools written by Dr. Allen Swei at Applied Biosystems for internal research and development use. Each metric is defined as follows (3):

a. *Length of Read (LOR):* Number of continuous bases that are called correctly (based on a known sequence) until a total error of 1.5% is reached.

b. *KB Q20 Length of Read (Q20 LOR):* Number of continuous bases in which the average KB quality score does not drop below 20 in a sliding window of 21 bases.

c. *KB Q20:* Total number of bases with a KB quality score of 20 or above.

d. *Peak Ratio:* Average ratio of the signals of the base-called peaks to the signals of any uncalled peaks underneath them.

e. *Peak Height Uniformity:* Deviation of the peak heights from the average local peak height on a per color basis.

f. *Total Signal:* Total average signal from all four bases.

Results

Due to the complexity of this study, we will focus discussion on the following three templates:

1. Templates with GC-rich regions
   a. Template #32 contains ~74 percent GC content (over ~980 bp)
2. Templates with GA Repeats or CT Repeats
   a. Template #143 contains a GA motif of >500 bases starting at approximately base 370.
   b. Template #580R contains a GA motif starting at approximately base 350.
   c. Template #513 contains ~46 repeats or ~100 base CT-repeat motif starting at approximately base 100.

3. Homopolymers
   a. Poly-G Region
   b. Template #773F contains a stretch of 30 guanines starting at base 705

Templates with GC-Rich Region

Template #32 is one of the hardest to sequence GC-rich templates in this study. We first sequenced all templates with BigDye® Terminator v3.1 kit at 1/16th × dilution to simulate customer-like conditions. Sequencing template #32 at this and even higher dilution levels (using less RR mix) results in quite noisy data because of very low signal levels (see Figure 3-1).

Because dGTP BigDye® Terminator v3.0 kit (containing dGTP) is formulated to address difficult G sequences, we sequenced template #32 with this chemistry next. Subsequent experiments with this chemistry resulted in a cleaner sequence with better signal; however, it does show G compressions (Figure 3-4). On the other hand, when template #32 is sequenced with the v3.1 and dGTP blend, instead of localized compressions, there is a rapid loss of resolution around base 215 that continues to worsen (Figure 3-5). The best result was actually obtained by sequencing template #32 with BigDye® Terminator v3.1 kit at the manufacturer's recommended protocol of 1× reaction. At this RR amount, the sequencing reaction generates enough signal to deliver a nicely resolved sequence with no mobility problems (Figure 3-6).

Polynucleotide-Repeats: Template #143, Template #580R, and Template #513

All three templates, when sequenced with BigDye® Terminator v3.1 kit chemistry, could not read through the problematic regions, even at non-diluted levels (Figure 3-7). Templates #143 and #580R are both examples of hard to sequence GA-repeat regions. The dGTP BigDye® Terminator v3.0 kit chemistry, this time, sequenced better than the v3.1 chemistry; it was able to sequence through problematic GA-repeat regions a bit more (Figure 3-8). The best option, in this case, turns out to be the v3.1 and dGTP chemistry blend. The blend was able to sequence through both templates #143 and #580R very effectively (Figures 3-9 and 3-10). Although we do see early loss of resolution with this blend on template #580R, it is slight and not as catastrophic as in template #32. The blend, thus, is still the best option for this template.

Template #513 contains a CT repeat that is one of the most difficult regions to sequence in this study. Like the repeat templates discussed earlier, this template resulted in a rapid stop not too far into the CT-repeat region when sequenced with the v3.1 chemistry (Figure 3-11). Since this

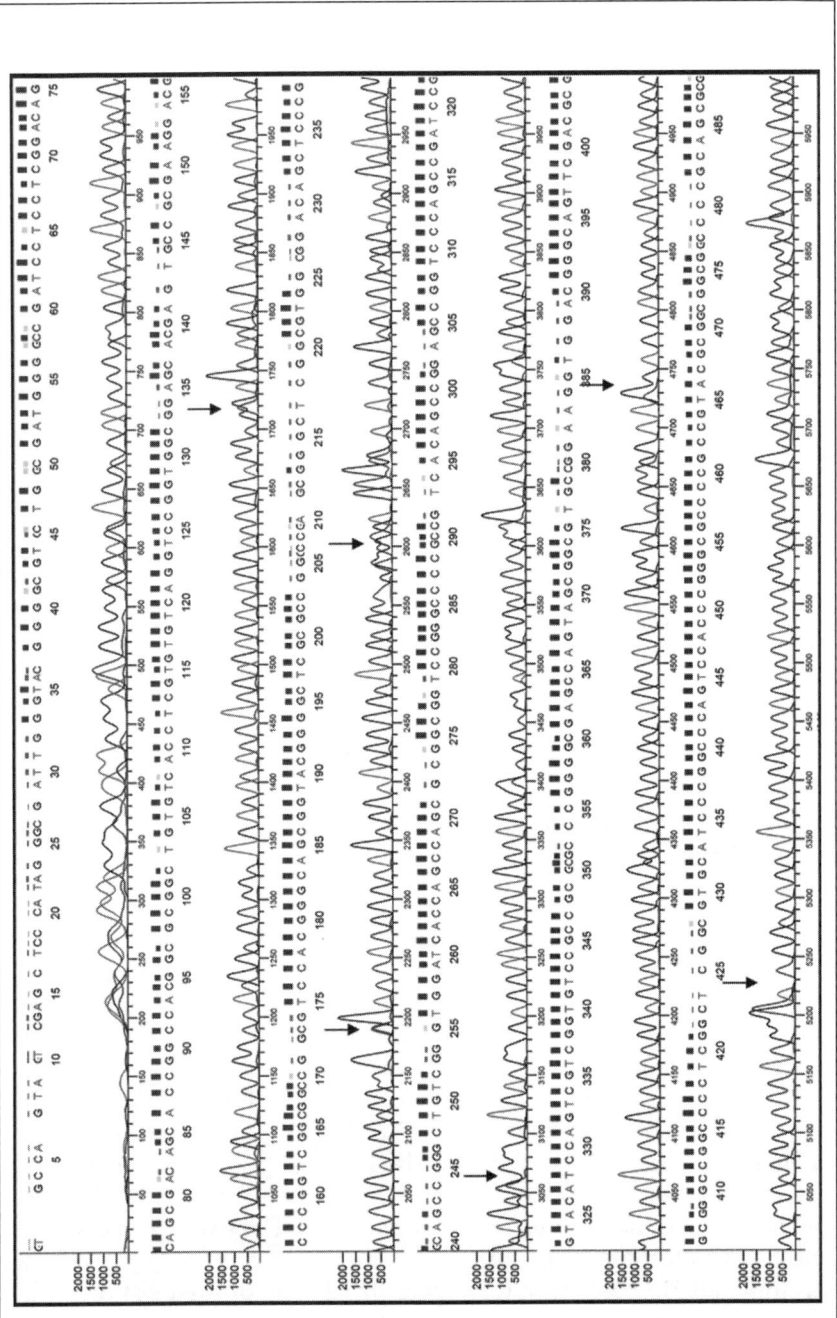

Figure 3-4. dGTP BigDye® v3.0 kit generates higher signal and cleaner sequence of this GC-rich template #32; however, it displays characteristic G compressions. *Arrows* indicate some areas showing G compressions. Notice the drop in quality scores (change from tall black bars to short gray bars) at these regions. G compression also manifests itself in overlapping of G and C bases and spacing issues, leading to shortened KB Q20 LOR. In this case, KBQ20 LOR (contiguous length of read of QV 20 or more) is only 423 bases.

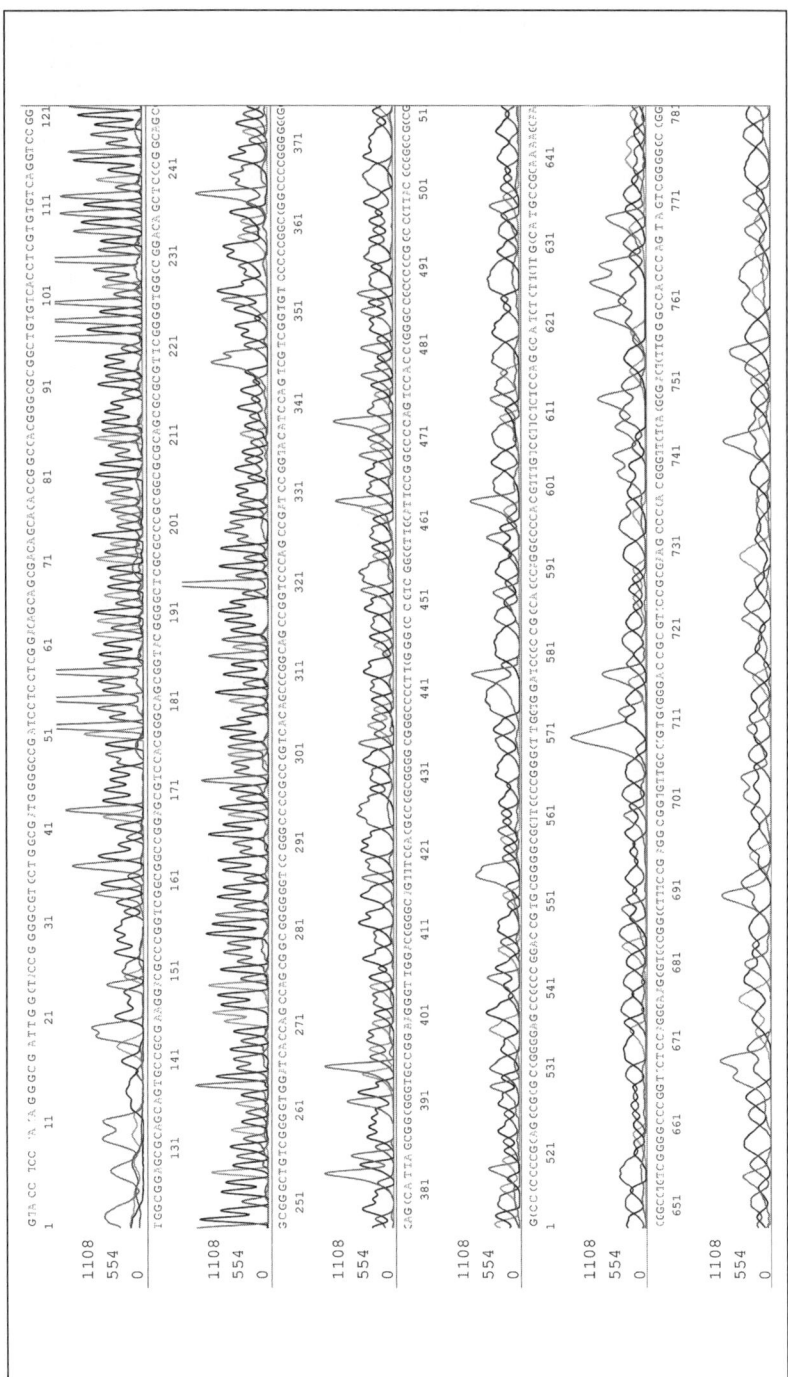

Figure 3-5. The BigDye® Terminator v3.1 kit and dGTP BigDye® Terminator v3.0 kit blend (75:25) generates good signal and no G-specific compressions in template #32, unlike those seen in **Figure 3-4**. However, the data show rapid loss of resolution at ~215 bp and continues to degenerate. Software could still basecall with good quality values (20 or more) until ~290 bp. KB Q20 LOR is 263 bases.

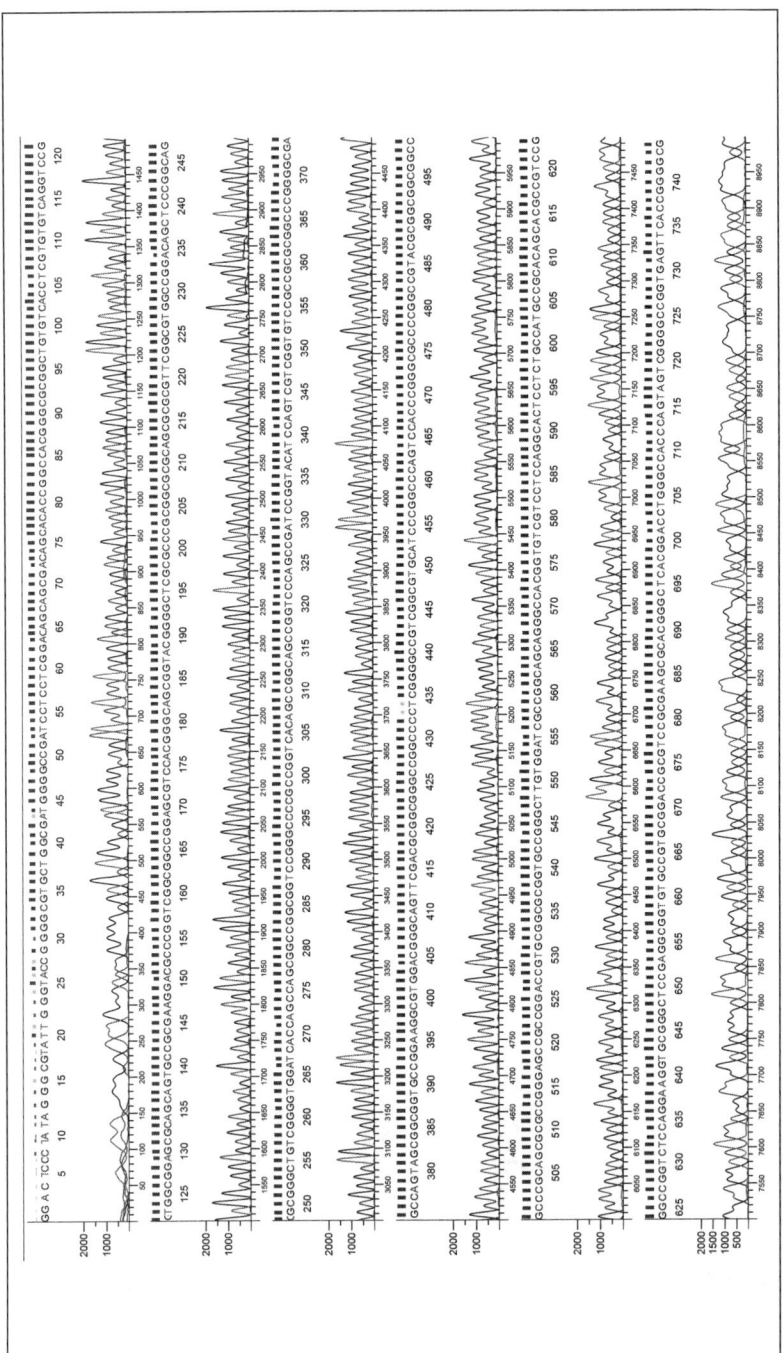

Figure 3-6. The best solution for sequencing this GC-rich template #32 is to simply not dilute. A 4 μL RR mix in a 10 μL reaction generates enough signal for a high-quality sequence with no compression or other mobility issues. In this sample, the sequence yields a KB Q20 LOR of 819 bp.

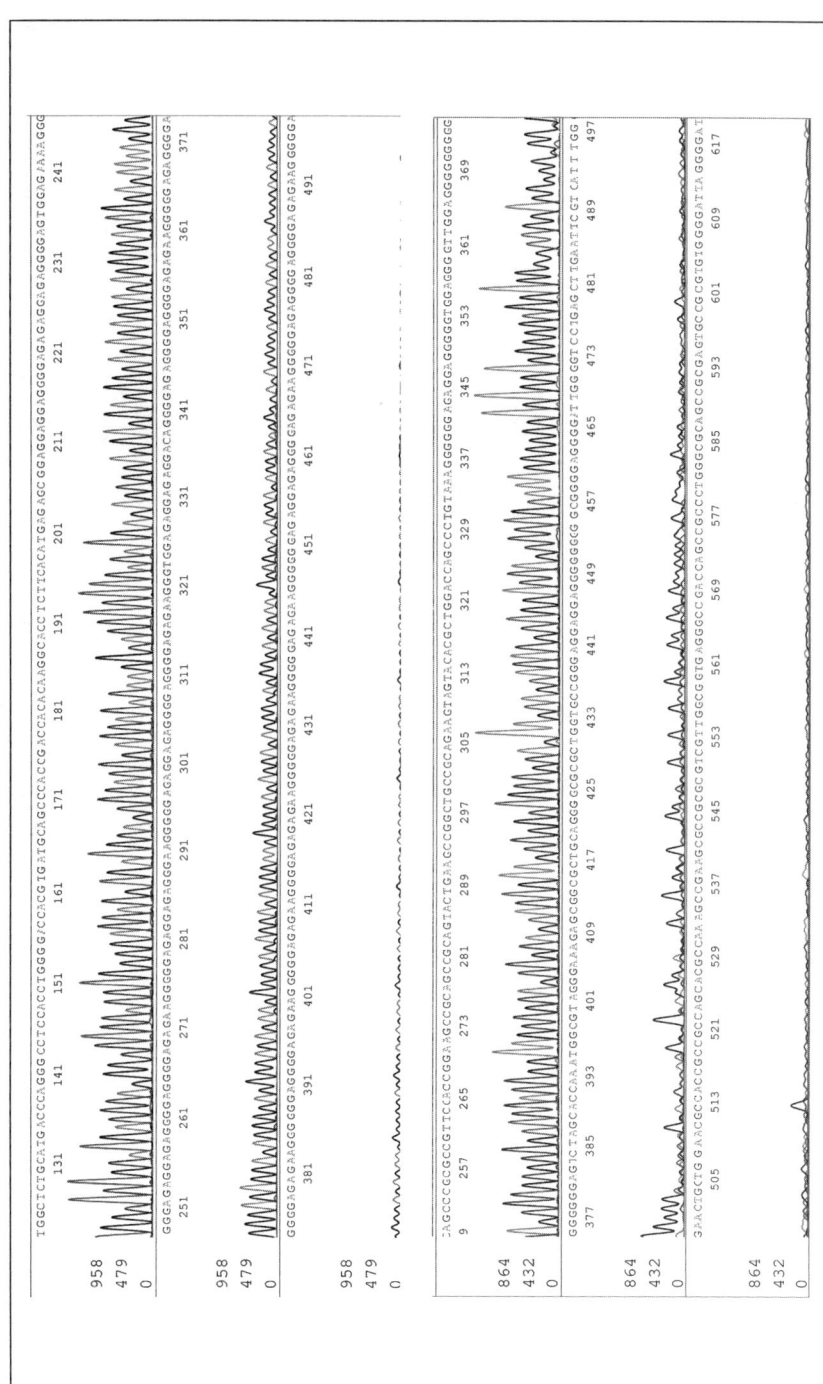

Figure 3-7. Dinucleotide repeat templates sequenced with BigDye® Terminator v3.1 kit showing early termination of extension product. Top panel: Template 143. KB Q20 LOR: 523bp. Bottom panel: Template 580R KB Q20 LOR: 360 bp.

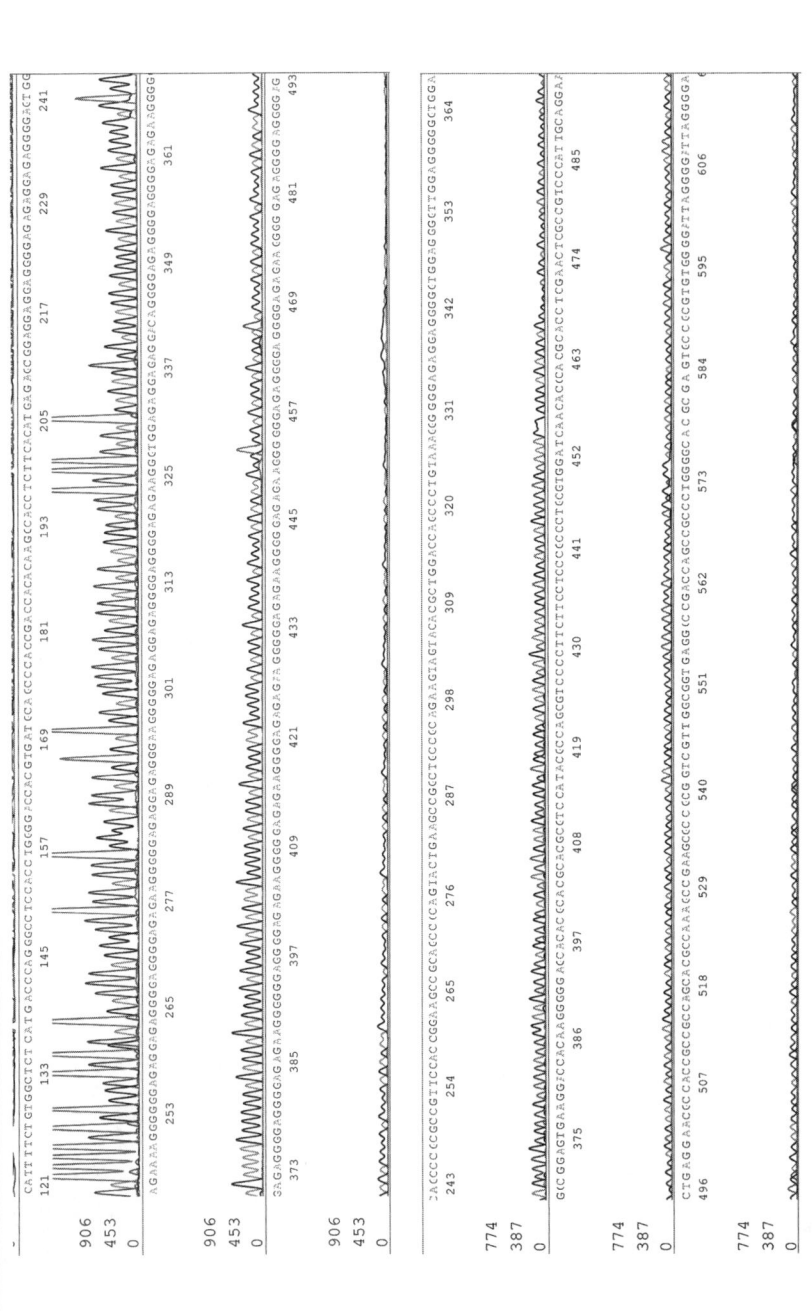

Figure 3-8. Dinucleotide repeat templates sequenced with dGTP BigDye® Terminator v3.0 kit showing longer extension product than with BigDye® Terminator v3.1 kit, but also gradual loss of signal. Top panel: Template 143, average raw signal: A (4639), C (1910), G (3229), T (2406). Bottom panel: Template 580R, average raw signal: A (860), C (975), G (776), T (1479).

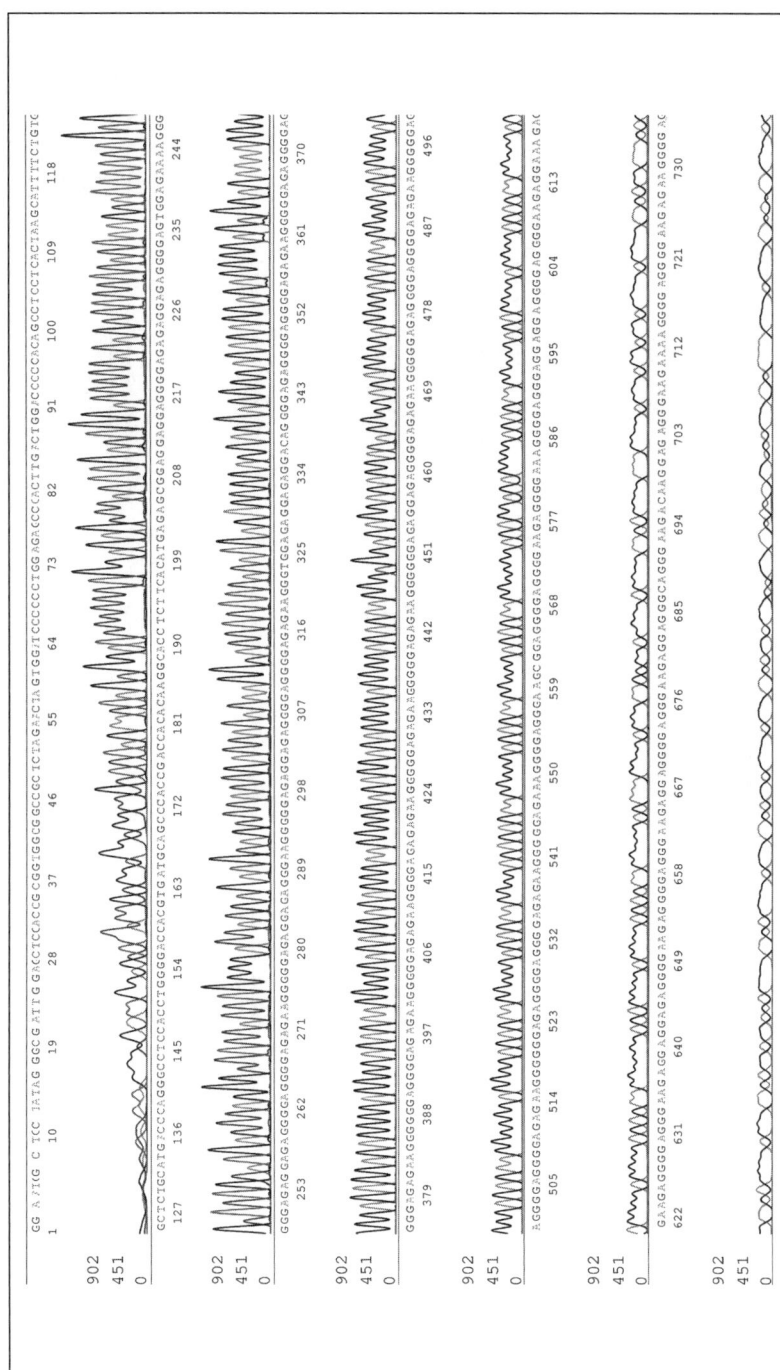

Figure 3-9. Blend of BigDye® Terminator v3.1 kit and dGTP BigDye® Terminator v3.0 kit sequenced template #143 very effectively through the >500 base G/A motif, achieving a KB Q20 LOR of 768. Although bases start to widen at about base 600, software can still basecall with good accuracy.

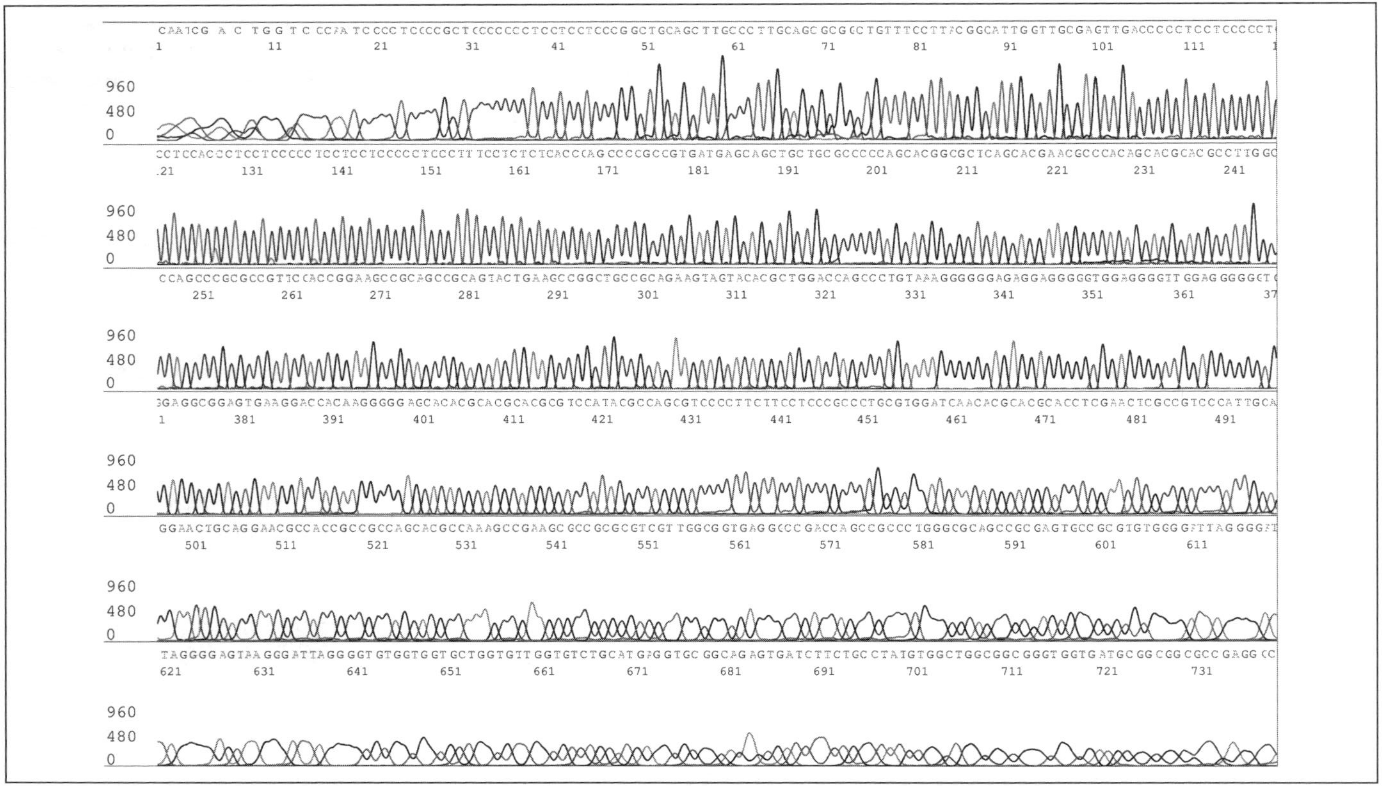

Figure 3-10. Blend of BigDye® Terminator v3.1 kit and dGTP BigDye® Terminator v3.0 kit can sequence through both the C/T and G/A motifs in template #580R with enough signal for high-quality basecalls (A:1602, C:1690, G:2115, T:1541). There is some early loss of resolution starting at ~600bp with this blend; however, it is not as catastrophic as seen in template 32, Figure 3-5. The blend is certainly the best option in this case.

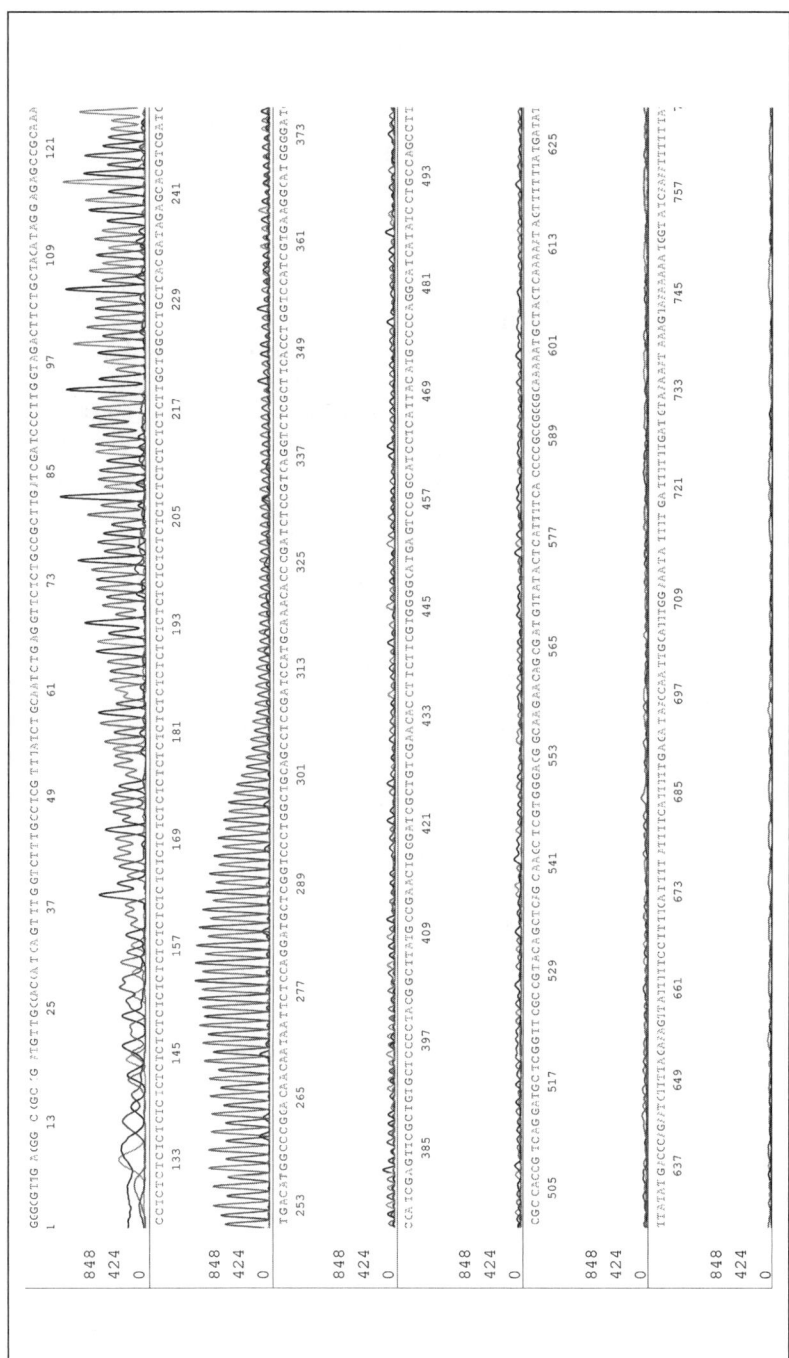

Figure 3-11. *C/T* dinucleotide repeat template, template #513, sequenced with BigDye® Terminator v3.1 kit showing early termination in C/T-repeat region. Although software tries to basecall peaks after the significant signal drop at around base 180, however, signals for individual bases after this point are so low that quality of basecalls is questionable.

is a CT-repeat difficult region, we did not expect either the dGTP or the blend to resolve this template. The best chemistry for this CT-repeat template is the Experimental Formulation F2. Formulation F2 sequenced through the entire 100 base CT repeat and the rest of the template efficiently without any loss of signal or resolution (Figure 3-12).

Homopolymer Regions: Template #773 and Others

As discussed earlier, homopolymer regions tend to result in slippage of bases in cycle sequencing. PCR tends to amplify the enzyme slippage problem, which leads to even worse sequencing results. Templates that are tested here are plasmids and thus did not go through a PCR step. We were able to sequence through these templates using two methods: changing primer proximity and formulating new chemistry mix.

Template #773 contains a stretch of homopolymer Gs starting around base 750. BigDye® Terminator v3.1 kit could not sequence through this spot when the original primer is 750 bases upstream. However, BigDye® Terminator v3.1 chemistry sequenced through the poly G region when the sequencing primer was redesigned to anneal ~125 bases away. The sequence did show some signal loss after the poly G region but still retain good quality bases. The dGTP chemistry, although it can sequence through with the new primer with the least signal loss, led to sequence that suffers worse slippage than BigDye® Terminator v3.1 kit chemistry (Figure 3-13).

Another homopolymer-containing template is a customer template with two homopolymer T regions. Because the first homopolymer region is close (~50 bases away from primer), BigDye v3.1 kit can sequence through; however, after the second homopolymer region, slippage becomes pronounced enough to cause base calling errors. We tested this template with experimental formulation F3 with good results. The F3 formulation sequenced the template with less slippage after both homopolymer regions, resulting in more high quality bases (Figure 3-14).

The F3 formulation also showed improved sequence quality in another homopolymer-containing template. The customer template with a 50-base homopolymer T region consistently showed slippage with either BigDye® Terminator v3.1 or the ABI dRhodamine kit chemistry. However, when using the F3 formulation, along with slightly modified cycling condition, the experimental chemistry generated data with considerably less slippage than the previously tested chemistries (Figure 3-15). The modified cycling condition is as follows:

96°C	hold for 1 min	
96°C	for 10 sec	
55°C	for 4 min	25 cycles
4°C	hold	

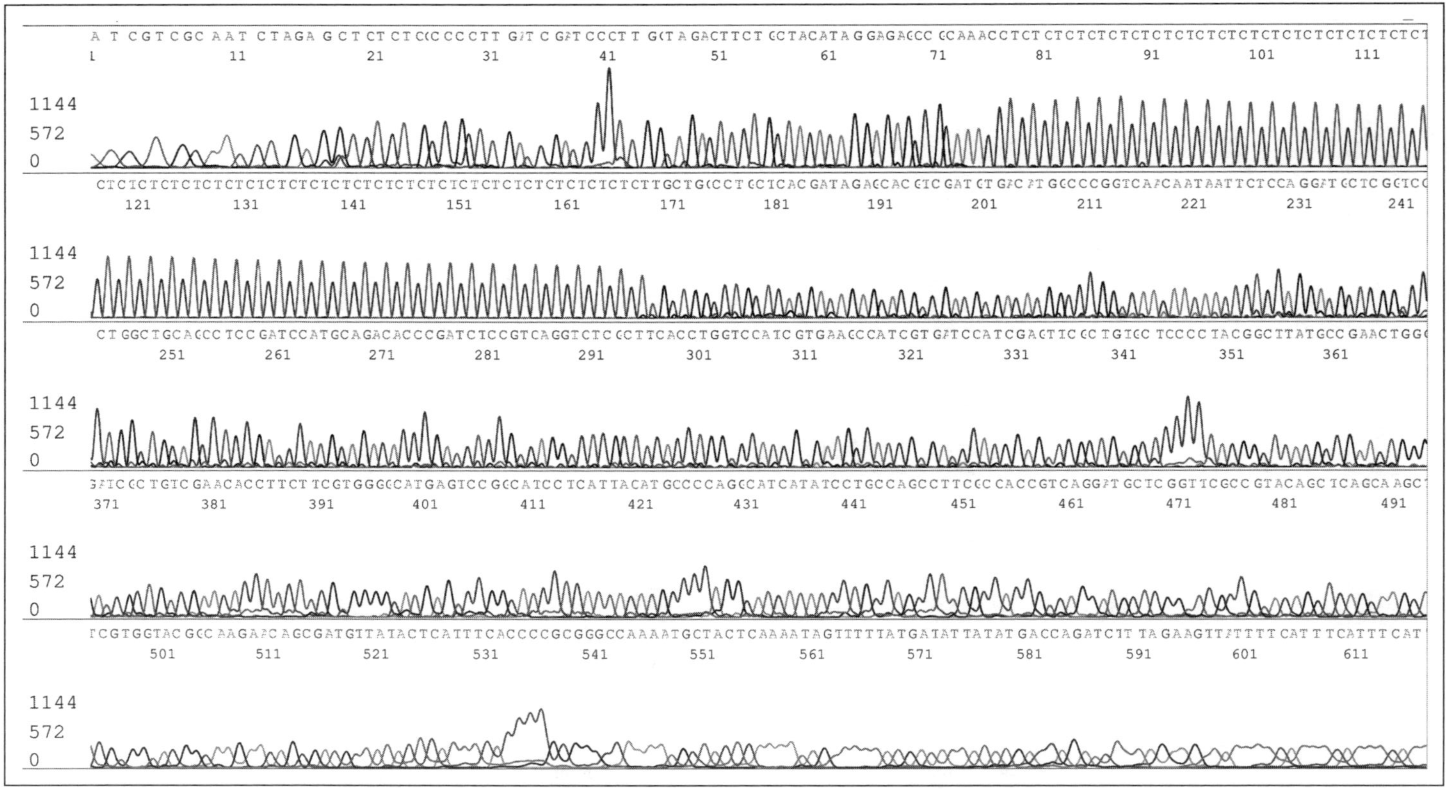

Figure 3-12. C/T dinucleotide repeat template, template #513, sequenced through the ~100bp C/T repeat region successfully with experimental formulation F2, achieving KB Q20 LOR of 674bp. Sequence following dinucleotide repeat region is slightly noisy but still gets good basecalls.

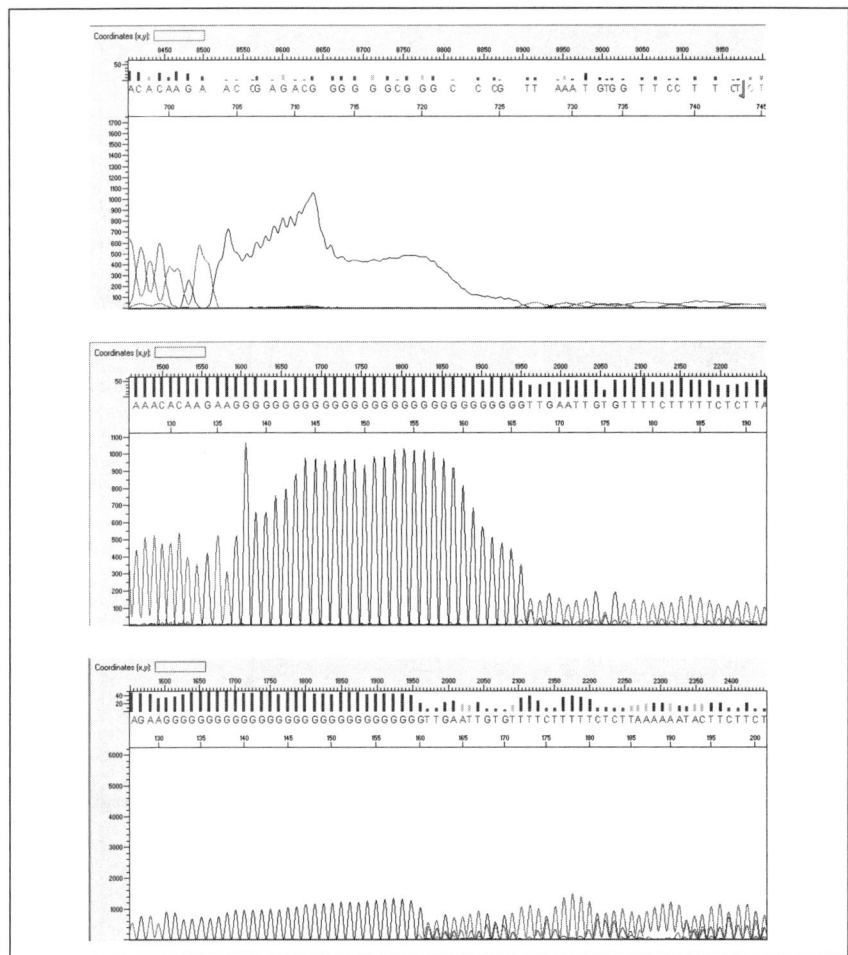

Figure 3-13. By moving primer closer to poly-G region, BigDye® Terminator v3.1 kit can now sequence through homopolymer G without loss of resolution or enzyme slippage. Top panel: Template 773 with BigDye® Terminator v3.1 kit. Primer is ~750 bp upstream. Middle panel: Template 773 with BigDye® Terminator v3.1 kit. New primer is now ~125 bp upstream. Bottom panel: Template 773 with Blend. New primer is now ~125 bp upstream.

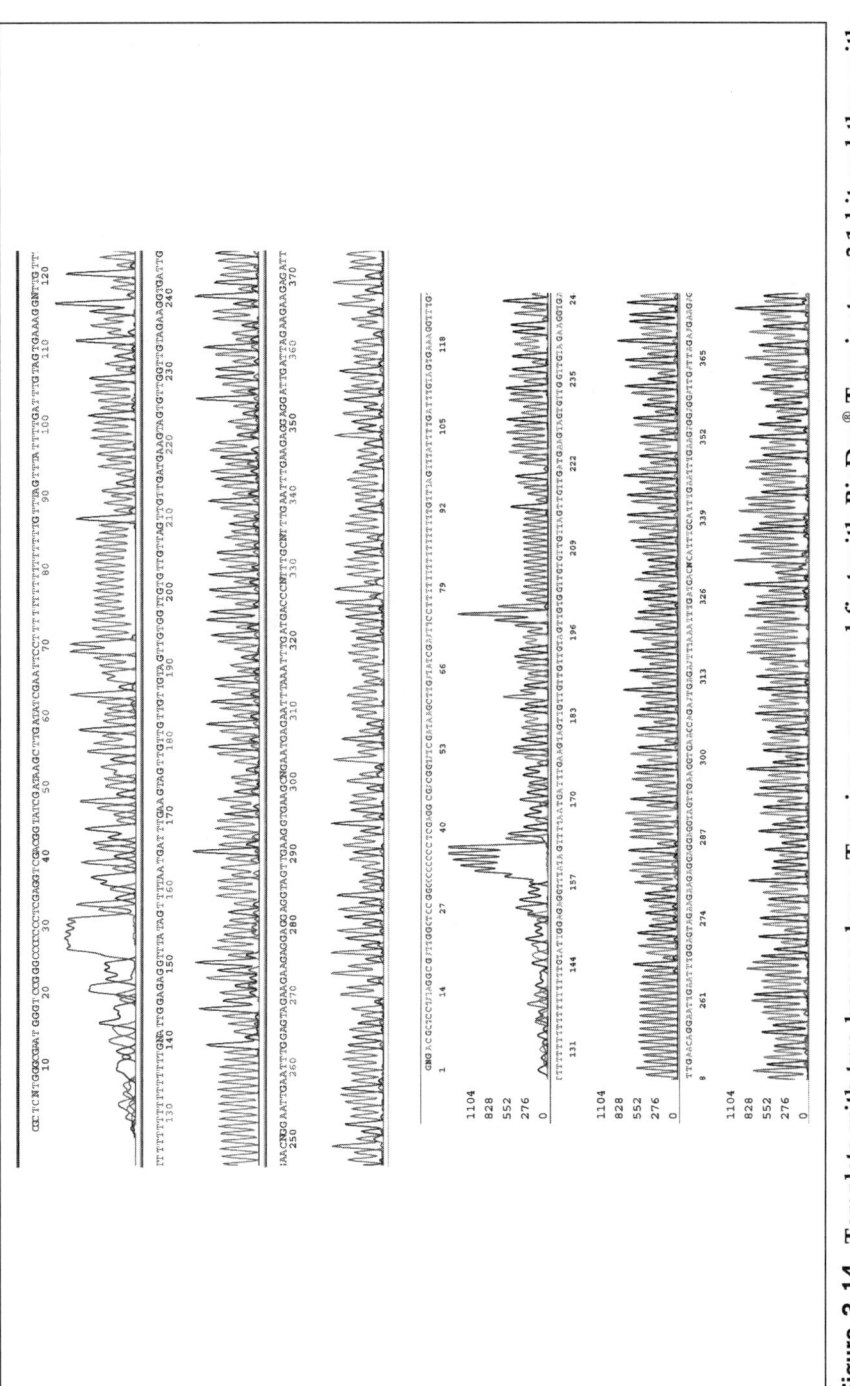

Figure 3-14. **Template with two homopolymer T regions, sequenced first with BigDye® Terminator v3.1 kit and then with Experimental Formulation F3.** F3 formulation sequenced this template cleaner than BigDye® Terminator v3.1 kit, with much less enzyme slippage and higher quality basecalls. Top panel: Template 126 with BigDye® Terminator v3.1. Bottom panel: Template 126 with Experimental Formulation F3.

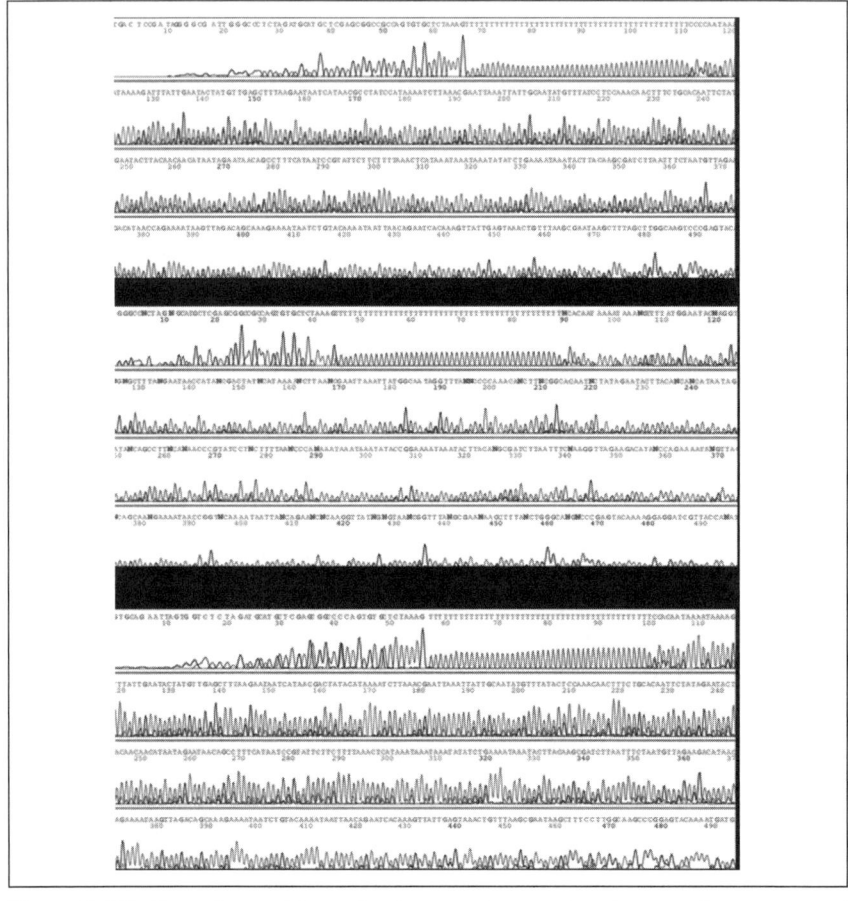

Figure 3-15. Template with ~50 bp homopolymer T region can be sequenced with considerably less slippage by modifying cycling conditions and using Experimental Formulation F3. Although one can still see secondary sequence in the Experimental Formulation F3 trace, it is basecalled with fewer N's and higher quality than either ABI PRISM® dRhodamine or BigDye® Terminator v3.1 traces.

Discussion

Although there is no single solution that enables sequencing through every sequence context, we were able to obtain satisfactory results for the most difficult templates with one or more of the experimental conditions tested. Below are our recommendations.

Do Not Dilute

For every difficult template that we tested, we found that as the RR mix amount becomes more dilute, the length of read goes down. As the reagent (enzymes, dinucleotides, dideoxynucleotides, etc.) amounts decrease, the sequencing reaction becomes less robust and therefore the overall sequence quality is lower. Since most of these difficult templates are hard to sequence due to high melting temperatures or long poly A/T tracts, the enzymes in RR mix may be falling off earlier or do not have sufficient binding accessibility. In a "full" sequencing reaction, there may be more enzymes that are still bound to difficult templates or that may have a better chance of binding than a diluted RR mix with less overall available enzyme and reagents. Therefore, we recommend using a more concentrated RR mix (less dilute) as a first step when troubleshooting in difficult-to-sequence regions.

Blending Works for Some Templates

The two cycle sequencing kits that we used are optimized to resolve different sequencing problems.

The dGTP BigDye® Terminator v3.0 kit is optimized for GT- and G-rich regions and uses dGTP. The BigDye® Terminator v3.1 Cycle Sequencing Kit has been optimized as the best all-around kit for all types of sequencing contexts, including GC-rich templates. This kit's major difference from the v3.0 kit is that it is formulated with dITP instead of dGTP. Although dITP-chemistry is very effective for routine GC-rich regions and obtaining long reads, it is known that in "difficult" G-region dITP, due to lower stability of bond formations, does not always incorporate well and will cause the reaction to stop (12). On the other hand, although dGTP incorporates better through these regions, this chemistry is susceptible to creating G compressions, meaning that although the sequencing reaction continues through the difficult region, the sequence is unreadable because of poor resolution. The G-compression problem may be caused by secondary structure formation of GC bases either at the terminal 3' end or within sequenced strands moving during electrophoresis at a similar rate and resulting in poorly resolved GC regions. According to studies done by Mills and Kramer, it is precisely dITP's less stable bond formations (2 hydrogen bonds vs. 3) with dCTP that help resolve G compressions when replacing dGTP with dITP in sequencing reactions (11). Prevalence of use by many core lab customers led us to hypothesize that blending the two kits could enable us to reap the benefits of both chemistries and read through the difficult regions while maintaining sequence resolution (7).

As our results show, blending these two different kits was effective for sequencing through some templates such as those with GA motifs. In

GC-rich regions, however, we observed the early loss of resolution when sequenced with this blended chemistry. We suspect that this is due to the sequencing enzyme pausing at the difficult G regions and incorporating a mixture of dGTP and dITP. Because these sequencing extension products have different mobilities, the peaks continue to broaden from this point onward. Thus, we recommend blending as a second troubleshooting step, particularly when faced with regions containing repeated, G-based motifs.

Primer Proximity

From previous studies (unpublished observations), we have anecdotal evidence showing that moving the sequencing primer closer to a homopolymer region as well as to other difficult to sequence regions gives the polymerase better access to the difficult region because the polymerase reaches the region before the two strands of the template start to renature. In our study, we moved the sequencing primer for a poly-G containing template from ~750 bases away to ~125 bases away. The new, more closely positioned sequencing primer successfully sequenced through the poly-G region without slippage or signal reduction. Because redesigning primers requires additional work, time, and money, we suggest that this tactic should be one of the last tools to try.

Conclusion

In this chapter, we have presented the results of several large studies of difficult to sequence motifs and the effect of sequencing reagents (both commercial and experimental) as well as the effect of dye-terminator dilution. Here are some suggestions (Table 3-1) that we have been able to generate for different categories of difficult motifs.

We are currently working to provide the experimental formulations tested in these studies as custom kits. There is no single answer for all of these difficult regions, but AB strives to continue to generate tools that can be useful in this effort.

Acknowledgments

I greatly appreciate the support and expertise provided by Diane Bond and Jackie Yen, the originators of this study. We would like to thank Dr. Geun-Sook Jeon and Dr. Shaheer Khan for providing experimental materials. We also would like to thank Dr. Robert Nutter and Dr. Peter Ma for critically reviewing this document and providing valuable feedback. Finally, this document would not have come to fruition without the editing expertise of Mignon Fogarty.

Table 3-1. Recommendations for sequencing of various difficult templates.

Context	Recommendation
polyA and poly C	BigDye® Terminator v3.1 kit
poly T	ABI PRISM® dRhodamine Terminator kit, F3
poly G	BigDye® Terminator v3.1 kit and primer design
AT and AC rich	BigDye® Terminator v3.1 kit
GA repeats	BigDye® Terminator v3.1 kit/dGTP BigDye® Terminator v3.0 kit BLEND
GA-rich	dGTP BigDye® Terminator v3.0 kit, BigDye® Terminator v3.1 kit/dGTP BigDye® Terminator v3.0 kit BLEND
GT-rich/repeats	BigDye® Terminator v3.1 kit/dGTP BigDye® Terminator v3.0 kit BLEND
CT repeats	F2, F3
CT-rich	BigDye® Terminator v3.1 kit
GC-rich	BigDye® Terminator v3.1 kit
GC repeats (CGG, GGC)	F2, F3
hairpin	BigDye® Terminator v3.1 kit, primer design
secondary sequence	modify cycling conditions

These recommendations are only suggestions for the best chance of successfully sequencing through respective difficult sequence contexts and are based on the results of the studies discussed in this chapter.

References

1. *Applied Biosystems 3730/3730xl DNA Analyzer User Guide: For use with Data Collection Software v.2.0.* 2003. User Guide. Part number 4347118 Rev. B. Foster City, CA: Applied Biosystems.
2. *BigDye® Terminator v3.1 Cycle Sequencing Kit Protocol.* 2002. Protocol booklet. Part number 4337035 Rev. A. Foster City, CA: Applied Biosystems.
3. Bond, D., Swei, A., Lee, K., et al. 2002. New Advances in DNA Sequencing Chemistry. Poster. Document number 106350. Foster City, CA: Applied Biosystems.
4. Bond, D., Yen, J., Yang, A., et al. Solutions for Sequencing Difficult Regions. 2006. Presentation at 2006 ABRF. Foster City, CA: Applied Biosystems.
5. Choi, J.S., Kim, J.S., Joe, C., et al. 1999. Improved cycle sequencing of GC-rich DNA template. *Exp Mol Med* 31: 20–24.
6. *Evaluating Capillary Electrophoresis Systems: Sequencing Difficult DNA Templates.* 2003. Fact Sheet DNA Sequencing. Document number 113158. Foster City, CA: Applied Biosystems.

7. Hawes, J.W., Knudtson, K.L., Escobar, H., et al. 2006. Evaluation of methods for sequence analysis of highly repetitive DNA templates. *J Biomol Tech* 17: 138–144.

8. Hengen, P.N. 1996. Cycle sequencing through GC-rich regions. *Trends Biochem Sci* 21: 33–34.

9. Kieleczawa, J. 2006. Fundamentals of sequencing of difficult templates—an overview. *J Biomol Tech* 17: 207–217.

10. Kieleczawa, J. Sequencing of difficult DNA templates. In: Kieleczawa, J. ed. *DNA Sequencing: Optimizing the Process and Analysis.* Sudbury, MA: Jones and Bartlett; 2005: 27–34.

11. Mills, D., and Kramer F.R. 1979. Structure-independent nucleotide sequence analysis. *Proc Natl Acad Sci U S A* 76: 2232–2235.

12. Spurgeon, S.L., and Brandis, J.W. New DNA sequencing enzymes. In: Kieleczawa, J. ed. *DNA Sequencing: Optimizing the Process and Analysis.* Sudbury, MA: Jones and Bartlett; 2005: 35–54.

4

Improving Sequence Results from Difficult Templates with Phi 29 DNA Polymerase and Nucleotide Analogs: the TempliPhi™ Sequence Resolver Kit

Haiguang Xiao and Carl W. Fuller
GE Healthcare, Piscataway, NJ

The chain-termination method of sequencing DNA is a relatively simple process that exploits the specific properties of DNA polymerase (the enzyme that copies DNA) combined with electrophoresis to precisely determine the size of DNA chains produced by the polymerase. For the method to work, all DNA synthesis must begin at the same sequence, a condition readily met by the use of a synthetic primer that anneals to cloned or otherwise amplified template DNA at a single unique site. The products of synthesis must be labeled, which can be achieved either by using a labeled primer or a labeled nucleotide including labeled chain terminators.

Chain terminators are nucleotides that are recognized by the polymerase as nucleotides and added to the 3′ end of the growing chain like nucleotides, but which terminate further synthesis because they lack a functional hydroxyl group at the 3′ position of the sugar (4). The terminators, in combination with polymerase, effectively convert sequence information into chain lengths because they terminate synthesis in a sequence-specific manner.

The most commonly used terminators are the 2′,3′-dideoxynucleotides in which a hydrogen atom replaces the normal hydroxyl group at the 3′ position of the deoxyribose. To function optimally, particularly when using labeled primers, the DNA chains should be synthesized continuously, stopping only when a chain terminator is added at the end of a growing chain. Stopping or pausing synthesis for any other cause will result in the production of a chain that may be interpreted incorrectly, adding "noise" to the experiment. One way to guard against this sort of noise in a sequencing experiment is to confine the label to the chain-terminating nucleotides themselves, which prevents the observation of chains not terminated by the incorporation of the sequence-specific chain terminator. This is the reason for the popularity of dye-terminator DNA sequencing methods; they are typically immune to many "false-stop" artifacts seen when the label is instead present in the primer or in the main chain of the newly synthesized DNA strand (2, 11).

The final element required for chain-termination DNA sequencing is the analytical method, which effectively measures the size of all the labeled DNA chains produced by the polymerase and nucleotides. Simple electrophoresis in polyacrylamide gels (in either capillary or slab-gel format) under denaturing conditions serves this purpose almost perfectly, as it readily resolves chains differing by a single added base up to sizes of 1000 nucleotides or more. Unfortunately, there are several situations under which the gels fail to resolve sequences adequately, where the DNA takes on a secondary structure that alters its rate of migration through the gel. When these structures form, migration of some fragments is faster than would be expected, resulting in bands that migrate too close to smaller fragments and giving the bands an appearance of being "compressed" together. Secondary-structure compression artifacts were well known even with the earliest gel separation methods, and appeared whenever GC-rich, inverted repeat sequences resulted in looped structures near the 3' terminus of the DNA. It was soon standard to replace the dGTP used in the sequencing process with an analog such as 7-deaza-dGTP (8) or deoxy-inosine triphosphate (dITP) (15), so that these secondary structures did not form and interfere with the experiment.

When dye-labeled dideoxynucleotides or dye-terminators were first tried, it was discovered that a second kind of interaction created a new class of "compression" artifacts. These are apparently caused by direct interaction of a dye (particularly rhodamine dyes) attached to the 3' end of a DNA and nearby bases, particularly guanines, within the chain (6). A number of nucleotide analogs were used to eliminate this kind of artifact, and the commercial cycle-sequencing kits all had dITP in place of dGTP for the four-color dye-terminator sequencing products (1, 14). This had the advantage of eliminating both kinds of compression artifacts: those caused by base-pairing as well as those caused by dye-binding to guanine or other bases. Because these commercial kits are the most popularly used sequencing products, accounting for more than 80% of all sequencing done during the last five to six years, many users are unaware that they are using nucleotide analogs at all or that sequence artifacts might have a different appearance with different sequencing methods.

The major advantage of dye-terminator sequencing—the overwhelming reason for its near-universal adoption for virtually all sequencing applications—is that only sequence-dependent terminations by a nucleotide terminator are actually labeled and detected in the experiment. Other ways of labeling DNA in sequencing experiments do not discriminate between terminator-mediated cessation of chain elongation and all other ways chain elongation can be slowed or stopped. For a simple example, consider what happens when using dye-labeled primer if the polymerase encounters the end of a template. At such an end, it simply stops making its complimentary copy of the template strand, regardless of the presence of a chain-terminating nucleotide. This situation is encountered when sequencing shorter polymerase chain reaction (PCR) products, or when adventitious PCR occurs during cycle sequencing with a labeled primer that finds two priming sites. Terminated fragments are produced in all the reactions (A, C, G, and T) so they are labeled with all four of the dye-primer colors. When the products are mixed and run on the sequencing instrument, a large peak, labeled with all four colors, is generated and the sequence is abruptly terminated. More commonly than reaching an end, template sequences are encountered that cause the polymerase to "pause" or stop regardless of the incorporation of a terminating nucleotide. Again, fragments are generated in all four reaction mixtures, thereby creating peaks labeled in all the colors. These nucleotide-independent terminations are a major cause of sequence ambiguities and artifacts when using dye primers for sequencing. They may be invisible when using dye terminators, but they still occur, and are commonly encountered as discussed below.

Before getting to the appearance of such artifacts in dye-terminator sequencing, there should be some understanding of the general trends in band intensities generated by sequencing methods. In any given dye-terminator reaction, 100% of the growing chains encounter the first base downstream from the priming site. A small fraction (e.g., 0.5%) of the chains terminates with the addition of a chain terminator at this very first site, so fewer chains reach the second nucleotide. Similarly, the number of chains that reach any nucleotide in the chain steadily decreases with the distance from the priming site. Thus, the intensity of bands (proportional to the number of chains terminated) steadily decreases the farther the sequence goes from the priming site. Because the nucleotide concentrations are adjusted to give sequences of 1000 or more nucleotides, typically the intensity of peaks is fairly steady within a neighborhood of about 50 bases in the final result, despite some local variation in intensity that results from local sequence context (6).

Materials and Methods

All DNA sequences shown were obtained using the DYEnamic ET Terminator cycle sequencing kit (GE Healthcare), running the reaction products on a Prism model 3100 genetic analyzer (Applied Biosystems, Foster City, CA). The resulting trace files were analyzed using Phred to determine the read-length as the maximum number of contiguous nucleotides with an average quality score of at least 20 (5). Plasmid templates with normal nucleotide composition were prepared directly from frozen, stored glycerol stocks of bacterial cultures using the method recommended in the TempliPhi™ DNA amplification kit (9). Briefly, a 1 μL aliquot of the glycerol stock was diluted with 10 μL of distilled water, and then 0.5 μL of the dilution added to 5 μL of Sample Buffer from the kit. This was placed in a sealed tube, mixed and heated to 95°C for one minute, then chilled on ice. Then a mixture of 5 μL of Reaction Buffer and 0.2 μL of Enzyme Mix (also from the kit) was added, and the resulting mixture incubated at 30°C for 16 to 18 hours. This mixture contains the Phi 29 DNA polymerase, random-sequence hexamers and sufficient dNTPs for synthesis of about 2.5 μg of DNA. After incubation was complete, the mixture was heated to 65°C for 10 minutes to inactivate the enzymes and kept on ice. Yields were typically 1 to 2 μg of DNA as measured by fluorescent dye staining. For sequencing, 2 μL of the amplified product (0.2–0.4 μg) was used following standard procedures.

Plasmid templates with modified (dITP + dGTP) nucleotide composition were prepared in the same way but using the reagents provided in the TempliPhi Sequence Resolver DNA amplification kit. The volumes were slightly modified as follows: 1 μL of diluted glycerol stock was mixed with 4 μL of Sample Buffer and denatured; then 4.5 μL of Reaction Buffer and 0.5 μL of Enzyme Mix added. Incubation of the modified amplification reaction mixture was done at 10°C instead of the usual 30°C because the melting temperature of the hexamer primers with modified DNA is significantly lower. Cycle sequencing reactions, however, were done at the usual temperatures. DNA yields were again 1 to 2 μg, but quantification of the DNA by fluorescent dye methods must be calibrated using DNA having the same base composition, or using UV absorbance methods.

Results

While there have been many methods used to prepare the DNA for sequencing, one of the most convenient is isothermal, rolling-circle amplification with the DNA polymerase from bacteriophage Phi29. The reagents for this method are available in the TempliPhi DNA amplification kits (GE

Healthcare) and are designed to amplify circular DNA sequences directly from tiny samples of bacterial colonies, phage plaques, glycerol stocks or as little as 1 pg of purified plasmid DNA, producing microgram quantities of DNA for sequencing in a few hours. The amplification works using random-sequence hexamers to initiate polymerase amplification of any DNA present in the reaction mixture, and the Phi29 polymerase synthesizes DNA in a highly processive fashion, displacing the non-template strand as needed to yield millions of copies of the DNA. The reactions usually proceed until the supply of nucleoside triphosphates runs out, so the amount of DNA made is constant and independent of the amount present at the start (9, 13).

The TempliPhi Sequence Resolver Kit is a variation of the TempliPhi kit that works in essentially the same way. The difference is that the nucleoside triphosphates present in the reaction mixture are an optimized mixture of dGTP, dATP, dCTP dTTP, and deoxyinosine triphosphate (dITP). Inosine is the nucleoside with a base like guanine but without the 2 amino group. This results in amplified DNA wherein most of the dG nucleotides present in the original DNA are replaced by dI. Because the dI-dC base pair has only two hydrogen bonds compared with the three of a G-C base pair, it is expected that this DNA should have less stable secondary structure and lower melting temperature. By simple optical melting experiments, we have found that DNA made with the Resolver Kit has a melting temperature about 20°C lower than that of DNA amplified by the normal mixture of four nucleotides (17). Using this DNA as template for DNA sequencing by the dye-terminator method can have a profound effect at some points where sequencing can be quite difficult.

Figure 4-1A shows an unusual dye-terminator sequence where the peaks quite suddenly drop in intensity and vanish altogether within a span of about six nucleotides. Examination of the rest of the sequence reveals no apparent reason for this sudden falloff, which looks as if the polymerase simply encountered the "end" of the template. In this case, however, the template is a 4-kb plasmid with no breaks. Nevertheless, all of the chains terminate within this short stretch of sequence, and most of them are not labeled by incorporation of a chain terminating nucleotide and so are "invisible" to the sequencing instrument.

The known sequence of this template is drawn in Scheme 4-1A, which reveals a 23-nucleotide inverted repeat that can form a stem-loop structure with a 23-base, 100% G/C stem and a loop of 16 nucleotides. This structure is remarkably stable, and is estimated to have a melting temperature (T_m) of about 107°C in the buffer typically used for sequencing (18). It appears that the polymerase cannot "read through" this sequence and simply stops about halfway through the stem. However, this is only half the picture. The results obtained depend strongly on the presence of

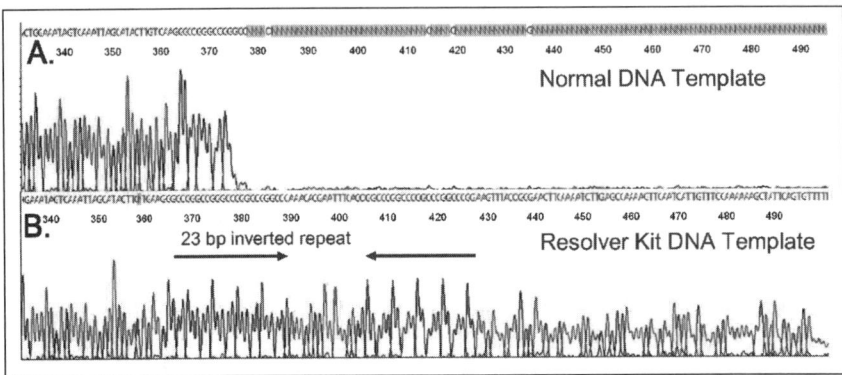

Figure 4-1. Sequences with GC-rich inverted repeats (see Plate 6 in the Color Addendum). (A) A portion of a sequence run on an ABI 3100 instrument using template DNA amplified with normal nucleotides and the TempliPhi DNA amplification kit. Similar results are obtained when sequencing reactions were carried out in the presence of DMSO, or betaine. (B) Results when the same template DNA is amplified using the TempliPhi Sequence Resolver Kit. A 23-nucleotide perfect inverted repeat composed exclusively of G and C bases is marked.

nucleotide analogs present in the sequencing reaction and the template strand as well. The growing chain in Scheme 4-1 (upper strand) is shown correctly with "I" (for inosine) in place of all "G" nucleosides. Keeping in mind that the polymerase-primer-template system is in dynamic equilibrium, structures like that shown as Scheme 4-1B are unstable, with the re-arranged structure (Scheme 4-1C) greatly favored over the usual primer-template structure because it has more of the favored G-C base pairs over the less favorable I-C base pairs. This structure with an unpaired 3′ end is either a dead end to DNA synthesis because it cannot be further extended or may be a site for initiation of a strand-switching event wherein the short 3′ single-stranded region becomes a primer for another nearby region of sequence (5). Note that strand switching with I-C base pairs may be unlikely at the 60°C elongation temperature.

At the temperatures used for typical dye-terminator DNA sequencing with dITP in place of dGTP, the re-arranged primer-template strands of Scheme 4-1C simply stop synthesis, giving results like that shown in Figure 4-1A. When dGTP is used instead of dITP, the polymerase typically does copy the entire stem-loop structure as might be expected, but both compression and strand-switching artifacts often make these results quite misleading or difficult to interpret. We have found that a better solution is to substitute inosine for guanosine in both the template and nascent strands. This is shown in Scheme 4-2 with the same template sequence. Presumably, with I-C base pairs, the stem-loop structure can still form, but with a greatly reduced stability. We have not directly measured the

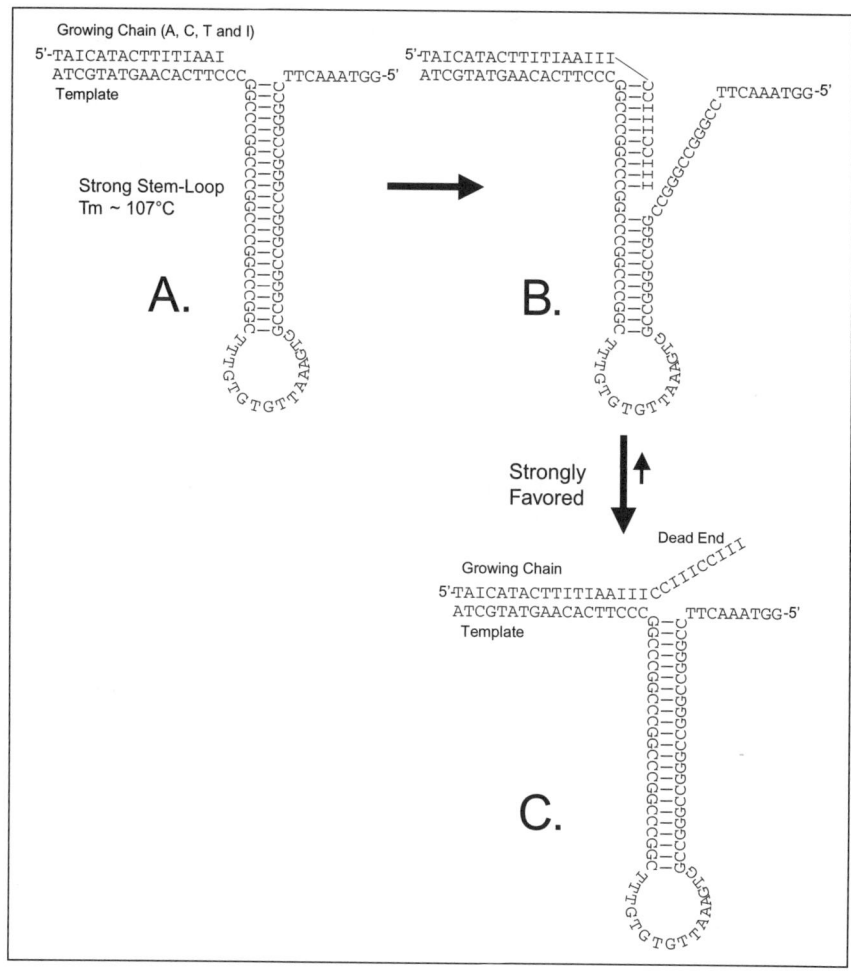

Scheme 4-1. During sequencing, the primer (upper) strand containing A, T, C and I (inosine) nucleotides is extended by DNA polymerase, usually remaining base-paired with the template (lower) strand. If the chain encounters an inverted-repeat sequence that can form a stem-loop structure (Figure 4-1), and the stem is GC-rich, particularly with a number of C nucleotides, the stem-loop duplex can be more stable than the template-nascent strand duplex since this strand has inosine (I) replacing the guanosine (G) nucleotides. Displaced from its template, the 3' end of the nascent strand can no longer be extended by polymerase.

stability of stem-loop structures like that shown in Scheme 4-2A, but we can estimate it roughly. First, we have measured the T_m of DNA having dI replacing dG in both strands generally. The melting temperature of a typical plasmid (50% G/C) was reduced from 95°C to 69°C when dI replaces all the dG nucleotides in the sequence. Second, a stem-loop

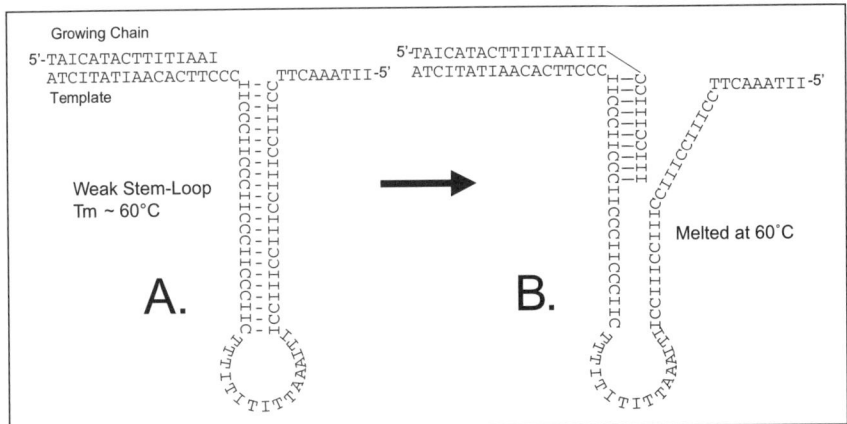

Scheme 4-2. The same template as shown in Scheme 4-1 is depicted with inosine (I) replacing all the guanosine (G) nucleotides as occurs when using the TempliPhi Sequence Resolver Kit. In this case, the stem-loop duplex is no more stable that the duplex formed by the template and nascent strand, so it melts as polymerization proceeds. The sequencing process traverses this region unimpeded.

structure like the one shown in Schemes 4-1 and 4-2 but substituting A for G and T for C has an estimated T_m of 61°C instead. Thus, it is reasonable to estimate that the T_m of the stem-loop structure would be reduced by more than 26°C (for 50% GC) but no more than 47°C (for changing 100% GC to 100% AT). Thus, the T_m of the stem-loop shown in Scheme 4-2A should be close to the temperature used for the elongation step in cycle sequencing (60°C). As the primer gets extended and the stem gets partially copied, it would simply melt, with no tendency to form unpaired "dead" 3′ ends on the nascent strand.

The result is an uninterrupted sequence as shown in Figure 4-1B. Here the entire inverted-repeat sequence can be seen with no sudden decrease in intensity and no ambiguities caused by compression or other artifacts. The template for this sequence was made by using the TempliPhi DNA amplification kit that uses Phi 29 DNA polymerase and random-sequence hexamer primers to amplify plasmid or other template DNAs. However, for this amplification dITP was substituted for most of the dGTP in the amplification step, resulting in a template that is largely, but not exclusively made of dI, dA, dC, and dT nucleotides with a minority of dG. Its T_m is typically an intermediate 75°C, about 20°C lower than a typical plasmid DNA.

The TempliPhi Sequence Resolver Kit takes advantage of these methods. First, the DNA is amplified using a mixture of dI, dG, dA, dT, and dC nucleotides, giving a product DNA that is much more readily sequenced than ordinary DNA. The advantage for stem-loop structures like that shown in Scheme 4-1 is quite clear and useful. It readily gives sequences of templates that cannot be done in any other way using dye terminator sequencing kits and it is much easier to use than schemes requiring the production of additional libraries (7) or switching to dye-primer or radiolabeled methods.

We have explored the use of this kit for a number of sequences by doing the following: First, we solicited a number of laboratories for templates defined as "difficult" solely by the contributing laboratory; we did not specify any sort of sequence features that they should have. This resulted in a collection of 91 "difficult" templates of various kinds. We then sequenced each of the 91 (and 5 controls and duplicates) using normal TempliPhi amplification and Sequence Resolver Kit amplification with either the DYEnamic™ ET terminator sequencing kits (GE Health-care) or the BigDye™ Version 3 (Applied Biosystems) sequencing kits. While results were similar for both sequencing kits, the ones shown were all obtained using the DYEnamic ET kits. We then tried to categorize the template sequences by comparing the results from normally amplified or Resolver Kit amplified templates and observed sequence content as shown in Table 4-1. The most common result (42 of the 91, 46%) was that both normal and Resolver Kit DNA gave good sequence results with little improvement for the modified template other than a modest (10%–20%) increase in read-length. (Also included in this category are templates where one or both sequences failed completely for unknown reasons.)

Of the remaining sequences, 37 (41%) were clearly improved by use of the Resolver Kit, 9 (10%) were about the same for both template prep-aration methods, and 3 (3%) gave worse sequence results. This last category included the most AT-rich sequences, including AT-rich repeats. Clearly, the TempliPhi Sequence Resolver Kit should be used with caution when encountering this category of sequences. A better strategy may simply be to use lower elongation temperatures for the cycle sequencing reactions along with the usual strategies of sequencing both strands using specific primers.

Of the sequences that were improved by preparing template with the Resolver Kit, 9 (Table 4-1) had perfect or nearly perfect inverted repeats of the sort diagrammed in Scheme 4-1. They all produced abrupt stops or reductions in sequence intensity giving virtually no sequence beyond the inverted repeat. When a sequence result is encountered with an abrupt stop at a run of 10 or more G and C bases, the Resolver Kit is likely to improve results dramatically like the results shown in Figure 4-1.

The next most common category had GC-rich sequences with exten-sive potential for secondary structures but the sequences did not stop abruptly, instead diminishing steadily over a span of 20 to 150 nucleotides, eventually becoming too weak for reliable basecalling. An example is

Table 4-1. Sequence results after using the Resolver Kit.

Description	Number of Templates	Comments
Strong inverted repeat >10 bp, >90% GC-rich; T_m > 85°C	9	Stopping suddenly and completely
GC- rich (>100 bp with > 70% G + C)	14	Gradual reduction in intensity over 20–100 bp
$(CT)_n$ repeats and poly pyrimidine tracts > 30 bp	5	Reduction in intensity over 10 or more bp
$(CA)_n$ repeats 50–250 bp	5	Repeats longer than 300 bp not resolved
$(GT)_n$ repeats > 20 bp	4	Many fully resolved
Poly C runs >15	3	Some improvement
Poly T runs (>15, variable)	2	Some improvement
Poly G runs > 20 bp	4	Little improvement
AT-rich sequences > 80% A + T, >200 bp	3	Somewhat worse with Resolver Kit
Uninformative results*	42	Both good, one or both failed
Duplicates, Controls	5	

* Because failures could have been accidental or amplification failures, they were not counted. When both the normal and TempliPhi Sequence Resolver Kit sequences were of high quality, the read-length with the Resolver Kit typically was 100–200 bp longer.

shown in Figure 4-2. The sequence result obtained with the Resolver Kit (Figure 4-2B) is clearly GC-rich, but while the sequence can form numerous secondary structures, they consist of either short (<10 bp) or interrupted inverted repeats with occasional A and T nucleotides. These results suggest that a variety of "dead-end" sequences form during the chain elongation step over a range of 50 to 150 nucleotides, prematurely starving the sequence of usable extension products. If you suspect your template is highly GC rich (these ranged from 65% to >80%) and the sequences appear to be weak or short or both, use of the Resolver Kit may well be an effective strategy.

Non-Repetitive Sequences Composed of Just Two Nucleotides

The collection of difficult templates included a number that had low-complexity sequences with long runs of just two of the four nucleotides.

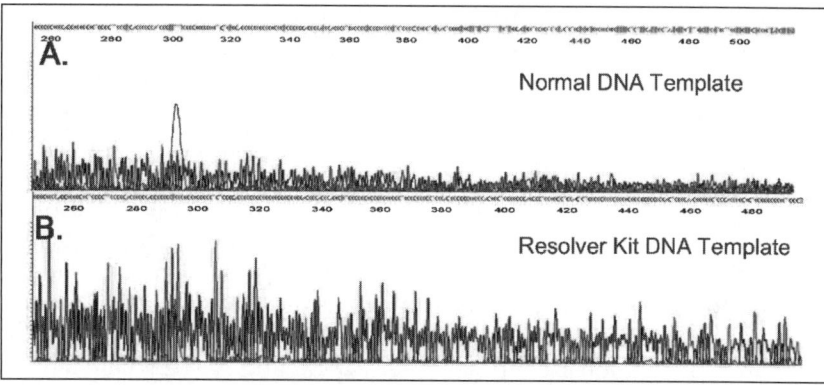

Figure 4-2. Extensive GC-rich sequences. (A) Results with normal amplified DNA. The intensity of bands decreases gradually, but more rapidly than they do on typical templates; not suddenly as seen in Figure 4-1. (B) The same sequence resolved in high quality with template containing inosine.

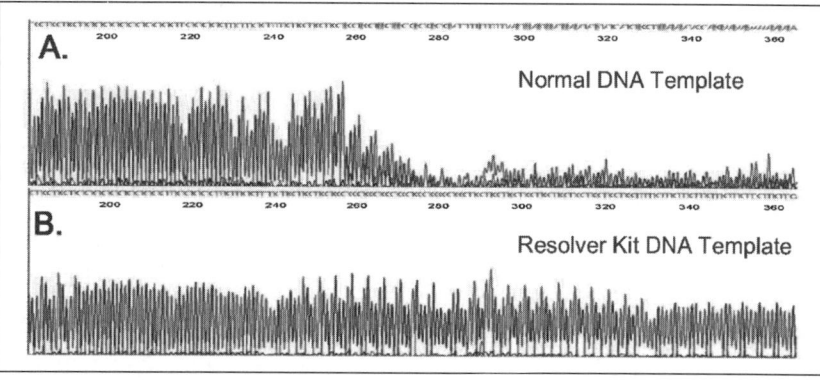

Figure 4-3. Polypyrimidine (CT) repeat sequences. (A) Results with normal amplified DNA. The intensity of bands decreases over a span of about 20 nucleotides, not as suddenly as seen in Figure 4-1. (B) The same sequence resolved in high quality with template containing inosine. DNA with one strand composed exclusively of pyrimidines and the other purines is known to have the potential to form triplex structures (see Scheme 4-3).

These included dinucleotide repeats, but typically also had a mixture of repeats of four, five, or more nucleotides as well.

For example, a C and T repeat sequence is shown in Figure 4-3. This polypyrimidine sequence is exclusively Cs and Ts for over 200 bp but lacks inverted repeats, and while some regions are C-rich, it cannot be described as GC-rich. As shown in Figure 4-3A, an abrupt decrease in sequence intensity and accuracy occurs over a span of about 20 bases. Keeping in mind that the template strand consists exclusively of As and Gs, it may not be too surprising that template prepared with the Resolver Kit, substituting dI for most of the dG gives a markedly different result (Figure 4-3B). The abrupt terminations seen cannot be the result of sequence structures like those of Scheme 4-1, but probably do result from some alternative secondary structure. It is interesting to speculate that a triple helix structure might be involved. Three-stranded, antiparallel structures form when a polypurine-polypyrimidine region with ordinary Watson-Crick base-paired structure is combined with a third polypurine strand which binds in the major groove forming "reverse Hoogsteen" hydrogen bonding, antiparallel with the original polypurine strand (10, 16). The structure will thus consist of base triplets, G*GC and A*AT, where the "*" indicates reverse Hoogsteen bonding (see Scheme 4-3 on page 102). For

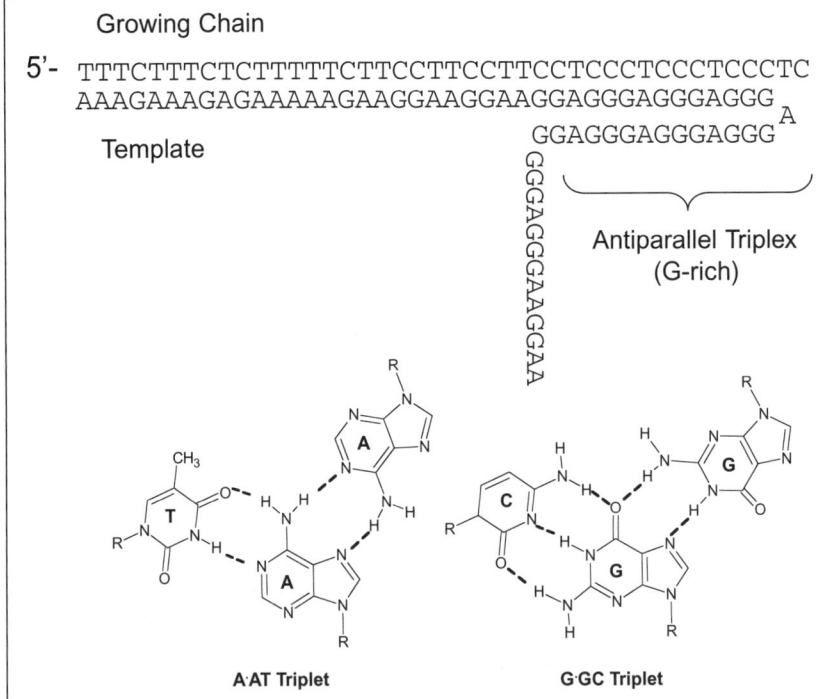

Scheme 4-3. When a polypurine template is sequenced, the nascent strand is made exclusively of pyrimidines. The difficulty in sequencing the DNA as shown in Figure 4-3 suggests that a secondary structure may be forming that impedes polymerization. Polypurine-polypyrimidine duplexes are known to be able to form antiparallel triple helix secondary structures given a third, polypurine strand, forming base triplets as shown (16). Such a folded structure would interfere with polymerase activity, and would likely be much less stable if the 2-amino group of G were not present to form two of the five hydrogen bonds of the G*GC triplet.

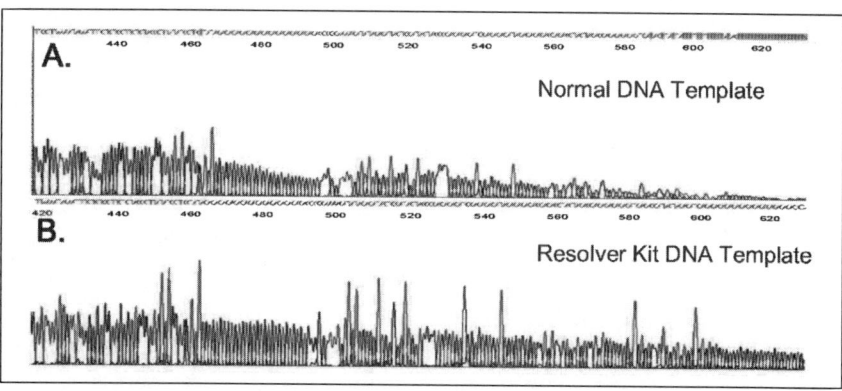

Figure 4-4. **Extended (CA) repeat sequences.** (A) Results with normal amplified DNA. While shorter stretches of CA repeats are readily sequenced, those longer than about 200 nucleotides can fail as shown with a steady decrease in peak heights. (B) The same sequence resolved in high quality with template containing inosine.

this structure to form, the polypurine template strand must have a sequence that can fold back on itself, pairing A with A and G with G and have the polypyrimidine strand present as well. From the schematic diagram, it is easy to speculate that the G*GC triplet will be disrupted if dI is substituted for dG, eliminating the possibility of hydrogen bond formation through the 2-amino group of G. Unfortunately, this is only indirect evidence that a triple-stranded structure can interfere with DNA sequencing. Other secondary structures such as G-quartet motifs could also be considered. As mentioned below, other polypyrimidine sequences have proven to be problematic, particularly C-rich ones.

The sequence shown in Figure 4-4 is composed almost exclusively of Cs and As but with some Ts and almost no Gs. As seen in the top panel of the figure, the sequence diminishes in intensity over a span of about 100 nucleotides, becoming too weak to read. With this and the other CA-rich sequences, it is clear that when longer than about 100 to 200 nucleotides, CA repeat sequences do not read well using normal dye-terminator methods. As shown in the lower panel, use of the Resolver Kit for template preparation relieves the difficulty. There is no obvious basis for a strong secondary structure in any of the CA-rich sequences encountered, but it appears that a collection of several shorter secondary structures may result in this kind of sequence difficulty. The template strand for these sequences is highly G-rich, and secondary structures in G-rich single strands are highly likely.

An additional instance of difficulty with a CA repeat sequence is shown in Figure 4-5. In this case, the sequence stops quite abruptly,

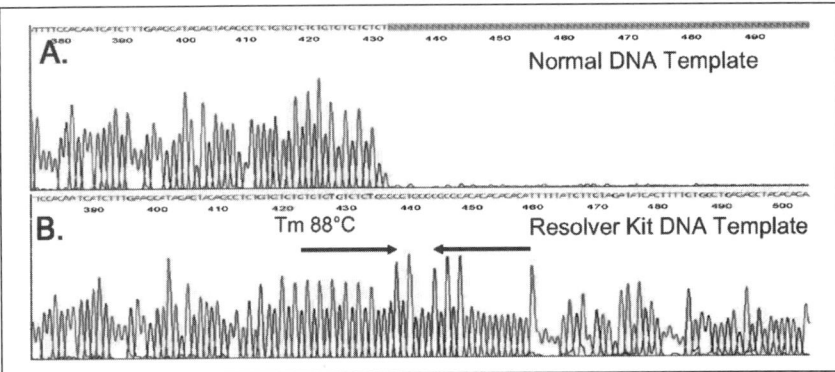

Figure 4-5. (GT) repeat sequences. (A) Results with normal amplified DNA. The sequence stops suddenly within a GT repeat region. The sequence in this region is only about 45% GC-rich. (B) The same sequence resolved in high quality with template containing inosine. The 16-bp inverted repeat that may be the reason the sequence stops suddenly is marked.

looking quite different from the case shown in the previous example. In fact, the sequence stops within a GT repeat sequence immediately preceding a CA repeat. Further inspection of the actual sequence reveals that the repeat sequence region is much shorter than the other problematic CA repeat sequences, and an inverted repeat sequence is evident, with the potential to form a 16 bp stem-loop structure. The stem is only about 60% GC-rich, and relatively short, but does have an estimated T_m of 88°C. Given the quite sudden drop in sequence intensity seen with normal sequencing methods (Figure 4-5A), it seems likely that this sequence is difficult primarily because of the stem-loop structure despite its relatively low GC content and not the result of extended CA-rich sequence as seen in Figure 4-4. An additional two templates in the collection had extremely long CA repeat sequences, both more than 400 bp long. These templates did not yield usable sequence with either normal or Resolver Kit amplification. For these cases, an alternative approach may be needed.

Homopolymer Repeat Sequences

While poly A and poly T sequences are usually sequenced reliably, difficulties can occasionally arise with homopolymer runs of 10 or more Ts, Cs, and Gs. The sequence collection used in this experiment had very few poly A or poly T repeats since these typically present little difficulty unless they exceed 100 bases in length. Figure 4-6 shows a sequence that includes 10 Ts in a run. As shown in Figure 4-6A, this sequence surprisingly appears to trigger some kind of difficulty both within and downstream of the short poly T sequence. More specifically, a second weaker sequence appears to be present along with the main sequence. While this can occur when two templates with different sequences are present in the same sequencing reaction mixture, the sequence is clear when the template is prepared using the Resolver Kit (Figure 4-6B). This suggests that some other mechanism, involving the G nucleotides in the template strand may generate this second, embedded sequence. This has been observed in rare cases where the 3′ end of the nascent strand actually switches template position, priming in a new place. It may take place as a concerted, intramolecular rearrangement (5) or during sequential denaturation and re-annealing cycles as are used for cycle sequencing and PCR. This kind of artifact is actually quite similar to strand "slippage" or "stuttering" where the 3′ end of a nascent strand can anneal to the wrong copy of a repeating sequence, resulting in a product that is either shorter or longer than a fully accurate copy. In this case, the source of the second sequence is not readily apparent, so the mechanism is difficult to pinpoint but fortunately it is fully resolved using the Resolver Kit for template preparation.

The final sequence shown (Figure 4-7) is one with a homopolymer of 29 Cs. It is not too surprising that the sequence is improved using the Resolver Kit, but it is not unambiguously resolved. In this case, the homopolymer is followed incorrectly by multiple Gs, again the apparent result of strand slipping. This polypyrimidine sequence is among the most difficult to resolve, capable of forming three-stranded structures as described

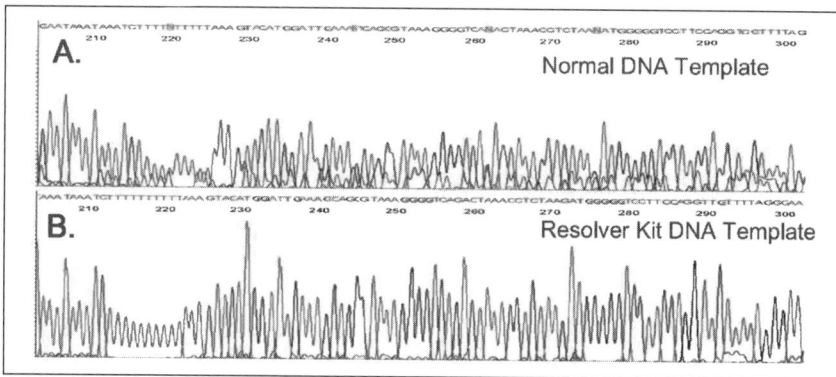

Figure 4-6. Homopolymer T sequences. (A) Results with normal amplified DNA. The sequence continues through and beyond the homopolymer region, but with an apparent second, weaker sequence along with the primary one, resulting in numerous errors and low quality. (B) The same sequence in high quality is obtained with template made with the Resolver Kit. It is not clear how such sequences might be improved by the use of the Resolver Kit, but several examples like this one were observed.

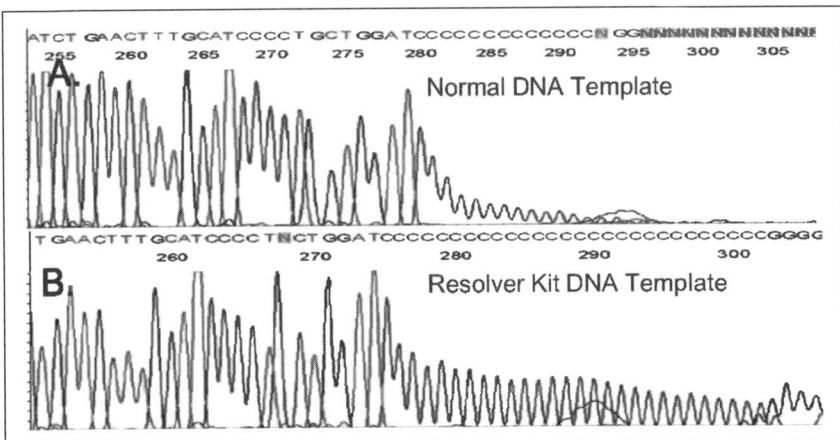

Figure 4-7. Homopolymer C sequences. (A) Results with normal composition DNA template. The sequence stops within a poly C region 29 bases long. (B) The same sequence resolving more of the poly C, but not extending with high quality beyond. This sequence, like the one shown in Scheme 4-3, could form an antiparallel triplex structure with the poly G strand folding back on itself. This structure may be less favorable when G is replaced by inosine.

in Scheme 4-3, strand switching and strong two-stranded secondary structures as well.

Discussion

The Sequence Resolver Kit is not a sequencing kit but instead is a DNA amplification kit that can be used to prepare DNA with altered nucleotide composition. Using the kit results in a template DNA with greatly reduced melting temperature (typically 20°C lower) and similarly reduced tendency to form secondary structures of several kinds. Amplification takes place in a simple, isothermal reaction mixture, requires only a few nanograms of starting DNA, and produces microgram quantities of template DNA for sequencing overnight. The DNA produced can be sequenced using ordinary dye-terminator and dye-primer sequencing kits, usually with improved, accurate results. Using a collection of 91 DNA template clones reported as difficult to sequence, the Resolver Kit consistently delivered improved results on the most problematic kinds of template DNAs that failed to yield unambiguous sequence in both directions. The only exception appears to be the AT-rich templates, which typically do not represent particular difficulties in most kinds of sequencing projects. The sequences most improved by use of the Resolver Kit are the ones with GC-rich sequences and ones with strong, GC-rich secondary structures, particularly those with several Gs in the nascent strand within the proximal portion of any stem-loop structures. In dye-terminator sequencing, these actually represent inosine nucleotides which compete poorly with guanosines in the template strand for forming intramolecular secondary structures. Also improved are sequences of polypyrimidine tracts that may form triplex secondary structures. A better understanding of why some sequences are hard to read using dye-terminator methods may lead to further improvements in the future. For example, additional nucleotide substitutions can be made, and these might improve sequencing performance, particularly on the longer poly C, poly G, and CA repeat sequences and highly AT-rich sequences that although rare, remain recalcitrant to sequencing by these methods. Methods that improve polypyrimidine sequencing and AT-rich sequencing may well find application when using sulfite-treated templates to search for 5-methyl cytosine in DNA for epigenetic studies (12).

Acknowledgments

We thank Robert Fulton of the Genome Sequencing Center, Washington University School of Medicine, for supplying us with collections of difficult template DNAs. We also thank Rebecca Deadman and Lucy Sha for initial evaluation of results, compilation of data from several laboratories and critical reading of the manuscript. Correspondence should be addressed to C.F. (carl.fuller@ge.com).

References

1. Bergot, J.B., Chakerian, V., Connell, C.R., et al. 1991. Spectrally Resolvable Rhodamine Dyes for Nucleic Acid Sequence Determination PCT. Patent Application WO1991005060.
2. Brummet, S.R., Ruan, C., Hosta, L.P., et al. 1997. [α-^{33}P] ddNTP Terminators—The New Standard for DNA Sequencing. *Amersham Life Sci Comments* 23: 1–7.
3. Ewing B., Hillier L., Wendl M.C., and Green P. 1998. Base-calling of automated sequencer traces using phred. I. Accuracy assessment. *Genome Res* 8: 175–185.
4. Fuller C.W. 2000. The Biochemistry of DNA Sequencing with DNA Polymerases. *J Clin Ligand Assay* 23: 249–255.
5. Lechner, R.L., Engler, M.J., and Richardson, C.C., 1983. Characterization of strand displacement synthesis catalyzed by bacteriophage T7 DNA polymerase. *J Biol Chem* 258: 11174–11184.
6. Lee, L.G., Connell, C.R., Woo, S.L., et al. 1992. DNA sequencing with dye-labeled terminators and T7 DNA polymerase: effect of dyes and dNTPs on incorporation of dye-terminators, and probability analysis of termination fragments. *Nucleic Acids Res* 20: 2471–2483.

7. McMurray A.A., Sulston J.E., and Quail M.A. 1998. Short-insert libraries as a method of problem solving in genome sequencing. *Genome Res* 8: 562–566.
8. Mizusawa S., Nishimura S., and Seela F. 1986 Improvement of the dideoxy chain termination method of DNA sequencing by use of deoxy-7-deazaguanosine triphosphate in place of dGTP. *Nucleic Acids Res* 14: 1319–1324.
9. Nelson J.R., Cai Y.C., Giesler T.L., et al. 2002. TempliPhi, Phi29 DNA polymerase based rolling circle amplification of templates for DNA sequencing. *BioTechniques* 32 (Suppl): S44–S47.
10. Plum, G.E., Pilch, D.S., Singleton, S.F., and Breslauer, K.J. 1995. Nucleic acid hybridization: triplex stabiliby and energetics. *Ann Rev Biophys Biomol Struct* 24: 319–350.
11. Prober J.M., Trainor G.L., Dam R.J., et al. 1987. A system for rapid DNA sequencing with fluorescent chain-terminating dideoxynucleotides. *Science* 238: 336–341.
12. Rakyan V.K., Hildmann T., Novik K.L., et al. 2004 DNA methylation profiling of the human major histocompatibility complex: a pilot study for the human epigenome project, *PLoS Biol* 2: 2170–2182.
13. Reagin M.J., Giesler T.L., Merla A.L., et al. 2003. A sequencing template preparation procedure that eliminates overnight cultures and DNA purification. *J Biomol Tech* 14: 143–148.
14. Rosenblum, B.B., Lee, L.G., Spurgeon, S.L., et al. 1997. New dye-labeled terminators for improved DNA sequencing patterns. *Nucleic Acids Res* 25: 4500–4504.
15. Tabor, S., and Richardson, C.C. 1987. DNA sequence analysis with a modified bacteriophage T7 DNA polymerase. *Proc Natl Acad Sci U S A* 84: 4767–4771.
16. Washbrook, E., and Fox, K.R. 1994. Comparison of antiparallel A·AT and T·AT triplets within an alternate strand DNA triple helix. *Nucleic Acids Res* 22: 3977–3982.
17. Xiao, H., and Hamilton, S. 2005. Multiply-Primed Amplification of Nucleic Acid Sequences. US patent application US20050239087.
18. Zuker, M. 2003. Mfold web server for nucleic acid folding and hybridization prediction. *Nucleic Acids Res* 31: 3406–3415.

5 Sequencing Through Various Secondary Structures: Detailed Studies of pDEST Vectors and Other DNA Templates with Hairpins

Tony Li, Mostafa Ait-Zahra, Paul Wu, and Jan Kieleczawa
Wyeth Research, Cambridge, MA

DNA sequencing through hairpin structures can be quite tricky. There are a few published reports using various protocols (4, 7, 10, 11) to sequence hairpins; however, none of them is robust enough to work most of the time. This is due to the complexity and the stability of the hairpin structures that prevent complete strand separation, necessary to obtain reliable sequence data. In fact, sometimes clean but false sequence readings are observed, as we show in this chapter. As more and more molecular biology applications involve experiments involving hairpin sequences, such as RNAi, shRNA reagents (3, 8, 9, 13, 14), etc., a robust protocol or strategy that consistently sequences through any hairpin structure will be of high value.

As a part of a standard practice, we at the DNA sequencing core group at Wyeth Research have completely re-sequenced almost every vector that was purchased from an outside vendor. On many occasions, the published sequence differs quite significantly from the "real" sequence; having the correct sequence helped to interpret data or explain unexpected results.

A few years ago, we experienced difficulties sequencing through Gateway Cassette of some of the Invitrogen destination vectors, such as pDEST8. Further study revealed the unexpected presence of a 27-base insertion in this vector (pDEST8, cat. #11804-010), which resulted in

DNA III: Dealing with Difficult Templates
Edited by Jan Kieleczawa
©2008 Jones and Bartlett Publishers

formation of a 24-base inverted repeat leading to a 24-base perfect hairpin structure. The presence of this hairpin caused many sequencing failures when trying to sequence though this structure. After a series of modifications of the standard sequencing protocol (1), we observed an unusual sequence pattern and developed a strategy to sequence through this hairpin, which later proved to be very successful in many other cases.

Materials and Methods

The original pDEST8 was purchased from Invitrogen (Carlsbad, CA) and was also prepared in-house using Qiagen® Plasmid Mini kit (cat # 12143, Qiagen, Valencia, CA); additional pDEST vectors were also purchased from Invitrogen. Other DNAs containing hairpin structures were prepared in-house using Qiagen prep method as described above. The standard DNA sequencing protocol was as described in (1) and basic modified protocol (including heat-denaturation) was as described in (10, 11). Other DNA sequencing modifications are described later in the text and in legends to figures and tables. The dye-terminator dilution factor in all experiments presented here was four times (2 µL of undiluted BigDye™ in 10 µL reaction).

The Not I restriction protocol, re-ligation, and all other procedures that lead to obtaining a plasmid without a hairpin were carried out using standard molecular protocols (15). Amplified sequencing reactions were prepared for runs as described in (10, 11) and analyzed on ABI3730 DNA analyzer (Applied Biosystems, Foster City, CA). The primary parameter in data analysis was read length ($Q > 20$) as described by Ewing (5, 6). If applicable, we also point out compression problems associated with dGTP chemistry (2).

Results and Discussion

The pDEST vector series are commonly used in molecular laboratories to accomplish high-throughput cloning (17). In this study, we concentrated on sequencing of pDEST8 vector but in addition we have sequenced quite a few other pDEST derivatives from Invitrogen, such as pDSET10, 14, 20, 24, and others. We found that the 24-base hairpin structure exists commonly in many of these vectors. The 27-base insertion happens to be flanked by a pair of Not I sites, which allowed for their easy excision by Not I restriction digest. The presence of this hairpin does not interfere with the cloning and sequencing of a cloned insert. It is likely that the plasmid vector sequence posted on the Invitrogen's Web site is stitched *in silico* and was not confirmed experimentally. Nevertheless, this pDEST8

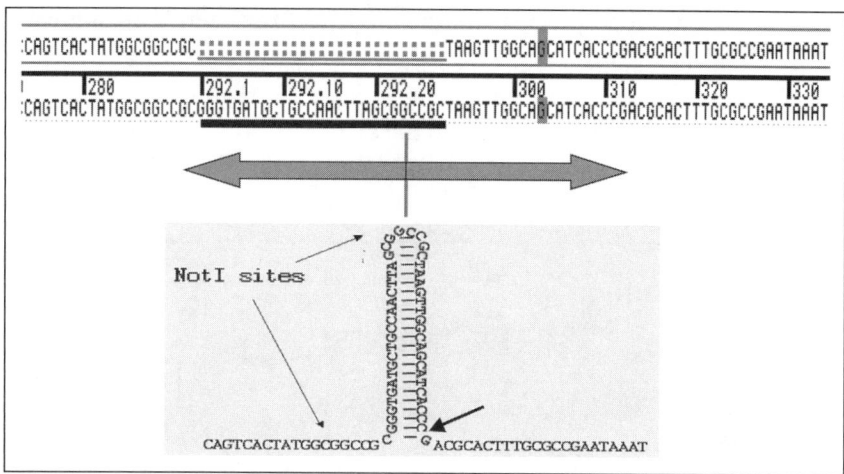

Figure 5-1. **Fragment of the pDEST8 vector sequence around MCS site.** The top part is from Invitrogen's Web site (http://www.invitogen.com; cat #11827011 and #11804010) and the bottom strand is the real sequence determined at Wyeth. The 27-base insertion (labeled with a horizontal bar leads to the formation of 24-base perfect hairpin as depicted in the lower part of this figure. The arrow at the bottom of the stem part of hairpin indicates 3′ base serving as a priming site for scenario #3 described below.

vector with 27-base insertion serves perfectly as a model of a difficult hairpin DNA template, and enticed us to perform a series of tests to develop a sequencing protocol that would efficiently result in clean read-through for this and similar structures (Figure 5-1) (see Plate 7 in the Color Addendum).

When a standard sequencing protocol (1) is used, the only sequence detected or observed continues to the beginning of the hairpin's stem and then abruptly stops. Using various modifications of the sequencing protocols (as described in the footnote to Table 5-1), two additional scenarios were observed. Thus, based on many trials we can distinguish three distinct sequencing outcomes:

1. Sequence reads clean until just a few bases after the Not I site (at the bottom of the stem) and then stops abruptly (Figure 5-2). This pattern occurs when the hairpin structure is not denatured under standard sequencing conditions and the DNA polymerase stops while encountering the double-stranded portion of the template.
2. As above, but after the clean stretch, a mixed peaks pattern is observed consisting of two or even three peaks in the same location, as seen in Figure 5-3.
3. Sequence is clean throughout the entire trace, but the peak intensity drops significantly a few bases after the Not I site (Figure 5-4).

Table 5-1. Comparison of various chemistries to sequence through pDEST8 hairpin.

Sequencing Protocol	Forward Primer (~470 bases to hairpin)		Reverse Primer (~160 bases to hairpin)	
	−HD	+HD	−HD	+HD
Control	546	743	172	600
+DMSO	512	593	174	472
+Betaine	507	553	163	416
+Reagent A	523	549	171	405
+Reagent C	524	482	244	217
+Reagent D	N/T	524	N/T	290
+Reagent E	N/T	530	N/T	344
+Reagent F	604	655	177	464
+Reagent G	479	526	173	250
+GC melt	484	509	168	377
dGTP V3.0	877	863	901	847
DB3.1 : dGTP 4 : 1	536	615	265	440
dGTP3.0 + Betaine	860	859	929	916
dGTP3.0 + A	858	851	919	899
dGTP3.0 + C	913	869	938	874
dGTP3.0 + G	922	920	931	885

Each data point (Q > 20 read length) is the average of triplicate reads. Reagents A through G (or just A–G) were from the Rx set of reagents purchased from Invitrogen (Carlsbad, CA). The GC-melt reagent is from BD Biosciences (San Jose, CA) and DMSO was from Sigma (St. Louis, MO). The 4 : 1 corresponds to 4 : 1 (v/v) mixture of BigDye™ terminator v3.1 and dGTP v3.0 (G. Grills, personal communication). The −HD refers to protocols without heat denaturation (standard sequencing protocol). Protocols with +HD include five minutes of heat denaturation at 98°C. All additives were present during this step. Each data point is the average of three independent readings and the standard deviation was in the range of 1% to 5% of the averages.

Note: Invitrogen discontinued selling Rx reagents in 2006 except for reagent A but they can be purchased as a special order. Anyone interested in purchasing these reagents needs to contact technical support at Invitrogen for details.

The detailed analysis of this last scenario reveals that the sequence reads "backwards" after the three Cs in the right side bottom of the hairpin's stem (indicated by arrows in Figures 5-1 and 5-4). It looks like the DNA strand breaks at this point and the 3' end of this fragment serves as the primer for extension of the sequence in the opposite direction. This

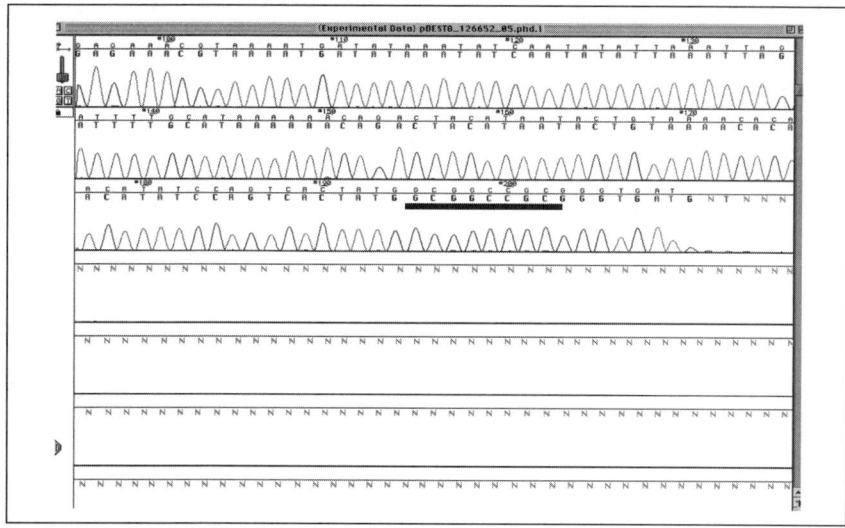

Figure 5-2. Abrupt termination of the sequence just few bases after the Not I site. The Not site is indicated with a horizontal bar.

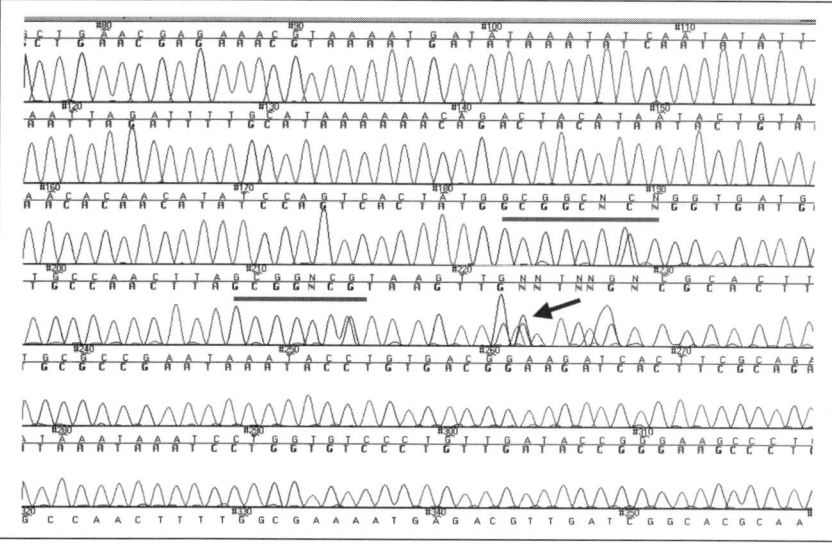

Figure 5-3. Very clean sequence till few bases after the Not I site followed by mixed base pattern. The Not I sites are pointed out with horizontal bars. Mixed bases are indicated by the *arrow*.

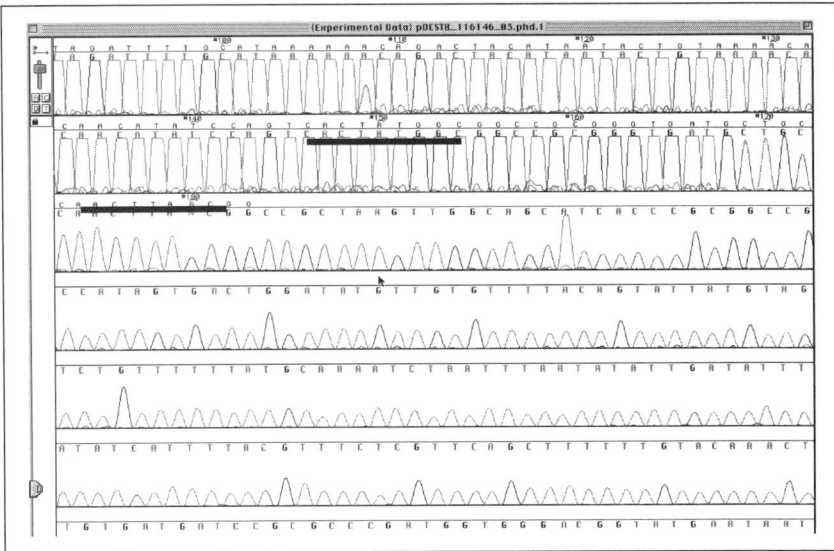

Figure 5-4. Very clean sequence pattern throughout the entire trace. Notice lower peak intensities just few bases after the first Not I site. Sequence is correct through the entire hairpin structure but instead of continuing in forward direction the sequence (GACGCA . . .) reads backwards creating another Not I site and reading the complementary strand after three Cs indicated by an *arrow*.

is further explained in Figure 5-5. Such an observation is not an isolated phenomenon, as a similar case was also observed in another, unrelated template (data not shown).

To interpret the scenario described in point #2 (on page 107), we believe that only partial denaturation of the hairpin structure occurs, resulting in stuttering of the DNA polymerase, which produces double or triple peaks.

To summarize the observations in this section, it is apparent that the outcome of the sequencing through this structure depends on the hairpin's denaturation state. When the hairpin is not denatured, the sequence reads cleanly until the beginning of the stem section of the hairpin and then it stops abruptly. If there is a mixed population of denatured and non-denatured hairpin structures, this results in a part clean sequence, followed by a mixed stretch. Finally, when the hairpin is fully denatured, as happens under extensive heat-denaturation conditions (11), this results in correct sequence (although small traces of secondary peaks are sometimes visible). To obtain a clean sequence through such structure, it is imperative that the hairpin is fully denatured, which is not possible when using standard sequencing conditions (1). Therefore, any effective step

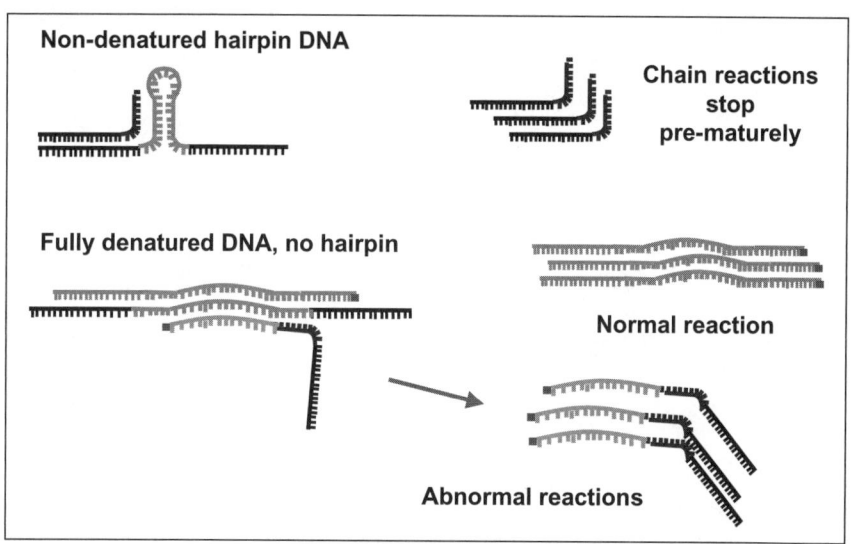

Figure 5-5. Suggested explanation for scenarios #1 and #2 described on page 107 (see Plate 8 in the Color Addendum). There are partially or fully denatured hairpin DNA templates under the standard (5) or modified (4) DNA sequencing conditions (in this case the modification includes just 5 minutes of heat denaturation at 98°C of the DNA template and primer). When the primer anneals to the fully denatured hairpin template, sequencing reactions will read cleanly through the hairpin region and generate the correct sequence (normal reaction). On the other hand, when primers anneal to the partially denatured DNA template, then the outcome will produce mixed population of terminated products resulting in overlapping peak signatures (abnormal sequencing reaction).

that leads to the denaturation or weakening of the hairpin structure needs to be employed before adding sequencing mix.

To obtain clean sequencing results, we tested many different conditions including modifications of existing protocols with and without additional additives. Results are shown in Table 5-1. The data presented in this table indicate that the dGTP3.0 chemistry is able to sequence through the hairpin region, but it suffers from compressed bases that are typically associated with this type of chemistry (2), as shown in Figure 5-6. Recently, GE Healthcare introduced a new kit (16; also see Chapter 4) that is very effective in sequencing through many types of difficult templates. We used this protocol to amplify the pDEST8 DNA; the resulting sequence is very clean and reveals the presence of the hairpin structure (Figure 5-7).

To confirm that the presence of the hairpin is the source of sequencing difficulties, we removed this hairpin through Not I digestion and re-ligation of the resulting DNA fragment. As expected (Figure 5-8), adding only a heat denaturation step resulted in clean sequence data. There is no

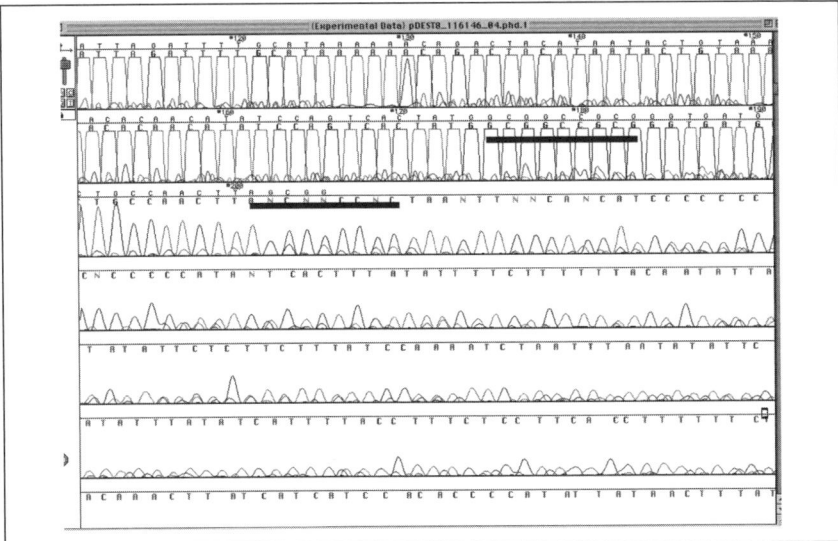

Figure 5-6. Sequencing through hairpin structure using heat-denaturation and dGTP 3.0 chemistry. Peak compressions are visible around Not I sites (labeled with horizontal bars) but the majority of the sequence is correct. Often the correct sequence can be deducted by combining data obtained from a regular (5) and dGTP 3.0 chemistries (12).

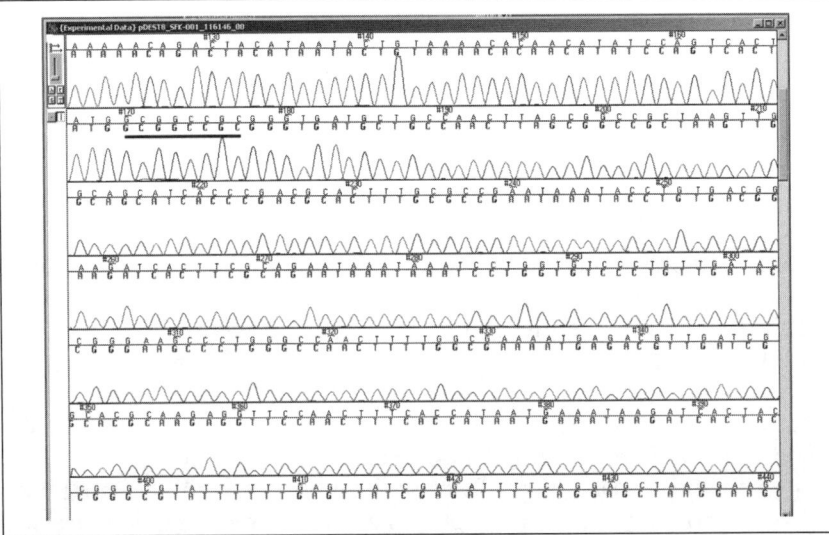

Figure 5-7. Clean read through the MCS region of pDEST8 following amplification using Sequence Resolver Kit. Position of Not I site is shown with horizontal bar.

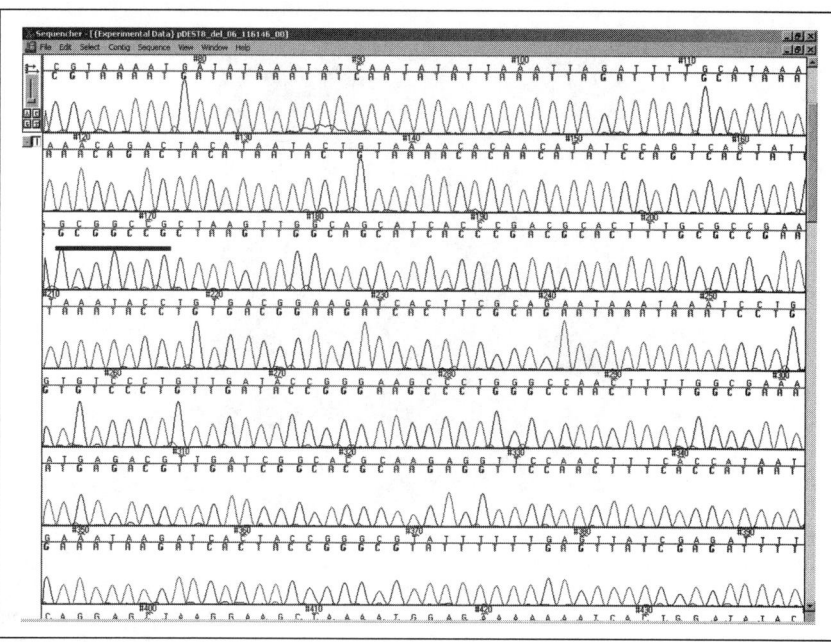

Figure 5-8. Good sequencing data after hairpin is removed from pDEST8 by Not I digestion. Notice the relatively even signal throughout the entire length of the sequencing trace. The Not I site is indicated with horizontal bar.

discrepancy between this and published sequence around the MCS region (alignment not shown).

The presence of the hairpin in this DNA template has a significant effect on the time needed to fully convert supercoiled DNA form into single-stranded form. Had this DNA been a "normal" template, five to six minutes of heating at 98°C in 10 mM tris/0.01 mM EDTA, pH 8.0 (= TE$_{sl}$) would have been sufficient to convert majority (about 75%) of supercoiled to single-stranded form (11; Chapter 1 in this text). Instead, at least 30 minutes under the same condition is needed to achieve a similar degree of conversion (Figure 5-9).

Re-Sequencing of Other pDEST Vectors

To make sure that pDEST8 is not a unique vector with an unreported hairpin structure, we completely re-sequenced a number of other pDEST-derived vectors from Invitrogen's collection (available at: http://www.Invitrogen.com). All re-sequenced vectors had numerous discrepancies compared to those posted on the Web site. In some cases, we could not

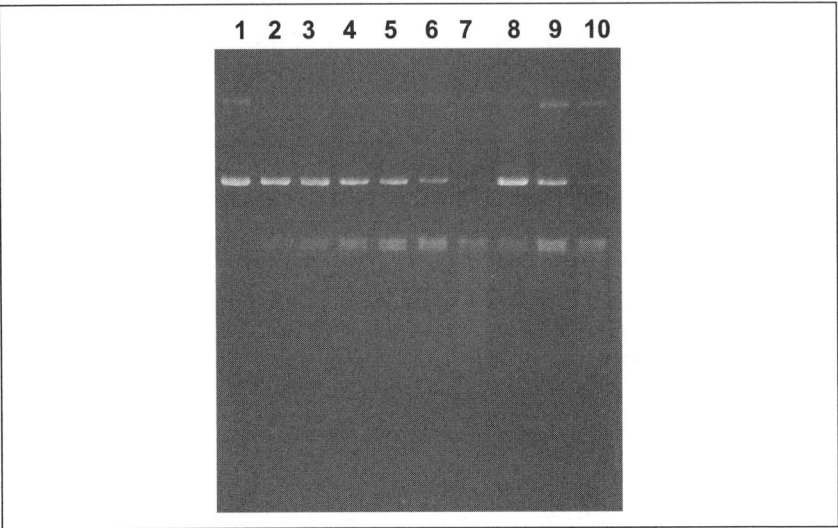

Figure 5-9. **Heat denaturation of pDEST8 DNA in TE$_{sl}$ and in water.** Aliquots of 200 ng of DNA were heat denatured for various periods of time either in TE$_{sl}$ (lanes 1–7) or in water (lanes 8–10) and entire samples were electrophoresed on 1% agarose gel as described in Chapter 1. Lane 1 = control, no heat denaturation. Lanes 2–7 were heat denatured for 2.5, 5, 10, 15, 20, and 30 minutes, respectively. Lanes 8–10 were heat denatured for 2.5, 5, and 15 minutes, respectively.

align (using simple alignment parameters) our sequences to the published ones, indicating that either significant deletion or insertion is present in a published sequence, preventing correct alignment. Table 5-2 shows comparison data between in-house sequenced and published pDEST vector sequences. Again, it is advisable that users check reference sequences in case their experimental data do not agree with the predicted behavior. It also would be valuable if companies offering vectors or other DNA-related products for sale indicate whether their sequence was confirmed experimentally or was stitched *in silico* from expected PCR-ed or restriction fragments.

Sequencing of Other Hairpin Structures

To expand the range of sequenced hairpins beyond pDEST vectors, we collected a number of plasmids (in the same vector backbone) containing 19-base hairpins with the same 7-base loop. These hairpin structures differ in GC content (from 38% to 53% GC) and the distribution of GC-pairs throughout the stem (Table 5-3). Each template was sequenced using eleven different protocols (described in the footnote to Table 5-3) using

Table 5-2. Comparison of published and experimentally confirmed sequences for pDEST vectors.

Number	pDEST Vector	Published Size (bp)	Experimental Size (bp)	Comments
1	pDEST8	6526	6554	1 inverted repeat/insertion-27 bp 3 small inserts/deletions 9 substitutions
2	pDEST10	6708	6716	5 inserts/deletions 8 substitutions/insertion-27 bp
3	pDEST14	6422	6446	3 inserts/deletions 5 substitutions/insertion-27 bp
4	pDEST15	7013	7009	2 inserts/deletions 5 substitutions
5	pDEST20	7066	7066	5 inserts/deletions 11 substitutions
6	pDEST24	6961	6958	1 inserts/deletions 5 substitutions
7	pET-DEST42	7440	7437	1 inserts/deletions 6 substitutions
8	pcDNA-DEST47	7952	7780	1 large deletion/insertion 245 bp 3 small inserts/deletions 8 substitutions
9	pMT-DEST48	5179	5179	1 large deletion/insertion 245 bp 3 small inserts/deletions 8 substitutions 9 substitutions
10	pcDNA-DEST53	7767	7767	

Remarks in the Comments column indicate various discrepancies between experimentally determined sequences and those posted on Invitrogen's Web site (http://www.invitrogen.com).

Table 5-3. Sequencing through 19-base hairpin expressed in terms of read length (Q > 20 values).

DNA Number	% GC	GC Distr	Tm (°C)	Meth Primer	1 -HD	2 +HD	3 +DMSO	4 +Bet	5 +A	6 +C	7 +F	8 +G	9 +GC	10 dGTP	11 4:1
1	47	1,4,5,7,8,12, 13,16,19	92	F	538	933	909	944	942	956	727	945	942	922	910
				R	0	404	800	918	895	665	376	506	852	378	718
2	42	1,2,5,8,11, 13,14,17	87	F	0	719	895	923	922	931	833	922	923	785	898
				R	467	468	632	942	920	671	473	673	699	458	811
3	47	2,8,9,10,12, 13,15,17,18	88	F	777	847	771	845	856	864	786	860	860	847	815
				R	967	955	955	957	959	952	952	969	962	960	951
4	37	1,5,6,7,9,10, 12	85	F	910	847	912	931	916	934	924	907	917	940	890
				R	481	496	598	910	796	654	575	531	894	738	701
5	42	3,4,5,10,12, 13,15,18	86	F	865	882	904	919	905	922	896	887	907	901	872
				R	N/T	N/T	N/T	N/T	N/T	N/T	N/T	N/T	N/T	N/T	N/T
6	53	2,4,5,6,7,8, 10,13,14,15	93	F	863	865	878	899	875	899	882	875	871	871	851
				R	835	821	884	902	870	895	880	880	858	584	862
7	42	2,3,5,8,9,12, 14,17	86	F	914	895	915	944	922	935	931	934	923	917	912
				R	882	878	922	937	932	929	915	912	906	923	905
8	47	6,8,10,11,12, 15,16,17,18	91	F	618	666	772	581	608	676	741	725	594	378	891
				R	340	467	644	926	879	891	414	881	932	548	755
9	47	2,4,6,8,9,12, 15,16,18	88	F	889	908	916	931	883	931	859	936	924	928	863
				R	937	877	931	938	885	904	869	935	741	802	838
10	37	3,6,10,12, 15,17,18	84	F	926	878	929	951	904	943	904	954	937	839	723
				R	983	974	980	978	969	985	976	975	974	967	933
11	47	1,3,5,9,10, 11,13,15,17	91	F	348	358	356	344	644	340	345	349	350	1036	1027
				R	170	189	175	288	159	183	167	178	181	1025	1014

Protocols 1 and 2 correspond to a standard (5) or modified (2, 4) sequencing methods. Protocols 2 to 11 included heat denaturation step and various additives (included in heat denaturation step). Reagents A, C, F, and G were from the Rx set of reagents purchased from Invitrogen. See Table 5-1 for other information on reagents and protocols. Each data point is the average of three independent reads and the standard deviation was in the range of 1% to 5% of the averages.

primers flanking the hairpin from both sides (forward and reverse primers). Our goal was to understand if there is a correlation between, for example, GC content, and how easy it is to get a clean read through a hairpin structure, as measured by Q > 20 read length values (5, 6). The analysis of data in Table 5-3 indicates that there is no such simple correlation between GC content or even the distribution of GC pairs throughout the stem and the length of clean sequence using standard sequencing protocol. On the other hand, it appears that if a melting temperature (Tm; as determined using calculations on Zucker and Markham's Web service) (18) of a hairpin structure exceeds 90°C, it is more difficult to obtain a clean sequence using even modified sequencing protocols. The 90°C threshold could be somewhat different for other types of hairpins, but at this point we do not have sufficient data to draw more accurate conclusions. In the near future, we are planning to embark on a much broader study to collect and analyze hundreds of various hairpin structures to try to establish a more precise correlation between, for example, Tm (or other thermodynamic parameter(s) and Q > 20 read length under different sequencing conditions.

Recently, we have sequenced 10 DNA templates with 19-base perfect hairpins in a different vector than those described in Table 5-3. Two of the hairpin constructs with Tm around 82° to 83°C sequenced easily using our standard protocol (10, 11). The other eight hairpin constructs with Tm around 94° to 96°C required an application of the protocol described in reference 16 to obtain clean read through.

Conclusions

Sequencing through hairpin structures continues to present strong challenges. Though a number of modifications have been published in the past few years, none seem to be universal enough to accommodate all hairpin variations (19–29 perfect matches or with 1–2 base mismatches). The data presented in this chapter indicate that there seems to be a broad correlation between Tm of a hairpin structure and the ease of sequencing through it using a sequencing protocol that includes the heat denaturation step. We recognize that a much larger set of data is needed to draw more definitive and general conclusions.

Applying various modifications to a standard sequencing protocol produces, in most cases, good quality and correct data while sequencing through hairpin structures as long as the calculated Tm value is below 90°C. If the melting temperature is above that threshold, more advanced protocols are needed to obtain clean read-through. Though using some additives is very convenient and efficient to sequence through many hairpins, they are no longer easily available on the market. On the other hand, one of the simplest, although slightly longer, protocols to sequence through most difficult hairpins is the application of the Sequence Resolver Kit from GE-Healthcare (16; this Kit was renamed recently from Sequence Finishing Kit to Sequence Resolver Kit; see also Chapter 4), although it requires about a 20-hour amplification step prior to the sequencing. We recommend that this protocol be applied mostly to the hairpins with Tm above 90°C, as the *a priori* application of Tm calculations (18) is very fast and helpful. In our view, the indiscriminate usage of this kit is not necessary as it leads to longer turnaround times and added expense.

References

1. *ABI PRISM® BigDye™ Terminator v3.1 Cycle Sequencing Kit. Protocol.* Part number 4337035 Rev. A. Applied Biosystems. Foster City, CA; 2002.
2. *ABI PRISM® BigDye™ Terminator v3.0 Ready Reaction Cycle Sequencing Kit. Protocol.* Part number 4390038 Rev. C. Applied Biosystems. Foster City, CA; 2002.
3. Berezikov, E., Thuemmler, F., van Laake, L.W., et al. 2006. Diversity of microRNAs in human and chimpanzee brain. *Nat Genet* 38: 1375–1377.
4. Ducat, D.C., Herrera, F.J., and Triezenberg, S.J. 2003. Overcoming obstacles in DNA sequencing of expression plasmids for short interfering RNAs. *BioTechniques* 34: 1140–1144.
5. Ewing, B., Hillier, L., Wendl, M.C., and Green, P. 1998. Base-calling of automated sequencer traces using phred. I. Accuracy assessment. *Genome Res* 8: 175–185.
6. Ewing, B., and Green, P. 1998. Base-calling of automated sequencer traces using Phred. II. Error probabilities. *Genome Res* 8: 186–194.
7. Esposito, D., Gillette, W., and Hartley, J.L. 2003. Blocking oligonucleotides improve sequencing through inverted repeats. *BioTechniques* 35: 914–920.
8. Fahlgren, N., Howell, M.D., Kasschau, K.D., et al. 2007. High-throughput sequencing of *Arabidopsis* microRNAs: evidence for frequent birth and death of MIRNA genes. *PLoS ONE* 2(e219): 1–14.
9. Kasschau, K.D., Fahlgren, N., Chapman, E.J., et al. 2007. Genome-wide profiling and analysis of *Arabidopsis* siRNA. *PLoS Biol* 5(e57): 479–493.
10. Kieleczawa, J. 2005. Simple modifications of the standard DNA sequencing protocol allow for sequencing through siRNA hairpins and other repeats. *J Biomol Tech* 16: 220–223.
11. Kieleczawa, J. 2006. Fundamentals of sequencing of difficult templates—an overview. *J Biomol Tech* 17: 207–217.
12. Lau, N.C., Seto, A.G., Kim, J., et al. 2006. Characterization of the piRNA complex from rat testes. *Science* 313: 363–367.
13. Rajagopalan, R., Vaucheret, H., Trejo, J., and Bartel, D. 2007. A diverse and evolutionarily fluid set of microRNAs in *Arabidopsis thaliana*. *Genes Dev* 20: 3407–3425.

14. Ruby, J.G., Jan, C., Player, Ch., Axtell, M., et al. 2006. Large-scale sequencing reveals 21U-RNAs and additional microRNAs and endogenous siRNAs in *C. elegans*. *Cell* 127: 1193–1207.

15. Sambrook, J., and Russell, D.W. *Molecular Cloning*, 3rd ed. Cold Spring Harbor, NY: Cold Spring Harbor Laboratory Press; 2001.

16. Sequence Finishing Kit. Product Code 25-6401-01. Piscataway, NJ: GE Healthcare; 2003.

17. Walhout, A.J., Temple, G.F., Brasch, M.A., et al. 2000. GATEWAY recombinational cloning: application to the cloning of large numbers of open reading frames or ORFeomes. *Methods Enzymol* 328: 575–592.

18. Zucker, M., and Markham, N. The DINAMelt server. Available at: http://www.Bioinfo.rpi.edu. Accessed November 3, 2007.

6

Transcriptional Sequencing as a Tool for Reading Difficult-to-Read Templates

Masanori Suzuki

RIKEN Genomic Sciences Center (GSC), RIKEN, and Yokohama City University, Yokohama, Japan

Complete decoding of the genomic information of an organism is one of the prerequisites for complete understanding of its entire function. In almost all cases, however, the genomic DNA contains various regions whose sequences are difficult to sequence using conventional methods. Moreover, a number of cDNAs also contain difficult-to-read stretches. Because the structural analysis of transcriptomes is based on cDNA sequencing, the presence of difficult-to-read regions in many of the cDNAs prevents us from fully understanding each of the transcriptomes. The cost and time for complete sequencing of many large genomes and comprehensive resolution of transcriptomes depends in part on the ability to sequence through such difficult regions.

Some major causes for the difficulty of sequencing are the existence of highly repetitive sequences, regions of high GC or AT content, areas of interstrand reannealing, inverted or direct repeats and their derivatives (i.e., hairpin stem-loops, long homopolymeric stretches, and some sequence motifs, causing band compression) (19). In particular, stable and complex highly ordered secondary structures with high GC contents greatly interfere with accurate sequencing. DNA polymerases (DNAPs) that are used for polymerase chain reaction (PCR) pause or break free from their templates at the GC-rich stretches located at the beginning of strong secondary structures, such as hairpins. This results in premature termination or base compressions (24) in the conventional cycle sequencing (CS) method. Many other artifacts also may be generated as a result of the formation of secondary structures in the single-stranded (ss)-DNA fragments by intramolecular Hoogsteen and/or reverse Hoogsteen base pairing (25). To overcome formation of secondary structures and undesired sequencing artifacts, several modifications and additives have been introduced in the sequencing reactions. Representative examples of such modifications and additives are the replacement of the guanine with a base analog (e.g., deoxyinosine 5′-triphosphate [dITP] [33] or 7-deaza-deoxyguanosine 5′-triphosphate [9, 23]) and addition of the Thermo-Fidelase (Fidelity Systems Inc., Gaithersburg, MD, USA), a hyperstable protein (archaebacterial DNA topoisomerase V), denaturant such as dimethylsulfoxide (DMSO) (29) and betaine (carboxymethyl) trimethyl-ammonium, *N,N,N*-trimethylglycine). By using ThermoFidelase and 2′-modified oligonucleotides (Fimers), the GC-rich genome of *Methanopyrus kandleri* was fully sequenced (31). The use of betaine has been reported to reduce band compressions in the sequencing of regions containing GC-rich base pairs, guanine stretches, or TGC-type repeats in several DNA templates (11, 12). Betaine can be applied in combination with DMSO for a uniform amplification of DNA with varying GC content and a CGG repeat region from the fragile X region (3).

However, these modifications and additives sometimes are ineffective for sequencing templates with GC-rich regions or highly repetitive sequences; thus, a sequencing strategy based on another principle is needed to confront these problems. Transcriptional sequencing (TS) adopts a different principle where double-stranded (ds) DNA is used as a template for the transcripts intercepted by chain terminators, instead of the ss-DNA used in CS. Because of this principle, TS can circumvent major sequencing difficulties caused by the high GC contents and the problematic secondary structures. Here an overview of TS is presented as an alternative or a complement to CS in the reading of the difficult-to-read sequences.

Materials and Methods

The CUGA sequencing kit (Nippon GeneTech Co. Ltd., Tokyo, Japan) was used for TS. The reactions were carried out according to the manufacturer's guidelines. Briefly, $4\,\mu L$ of RNase-free plasmid DNA containing a target DNA segment to be sequenced was mixed with $4\,\mu L$ of 5× reaction buffer, $1\,\mu L$ of $8\,mM$ $MnCl_2$, $2\,\mu L$ of rNTP/Termination mixture, and $8\,\mu L$ of ddH_2O. When some DNA region was found to be difficult-to-read by conventional dideoxy-chain termination sequencing, the target DNA region to be sequenced was amplified by PCR with the specific primers linked with T7 or T3 promoter and cloned into a TS vector such as pTS1. Freshly diluted $1\,\mu L$ of enzyme solution was added and mixed gently. Mixtures were incubated at 37°C for 60 minutes. Reaction products were purified and recovered with EtOH precipitation. Electrophoresis and detection were performed using an ABI PRISM 377 auto sequencer XL or 310 (Applied Biosystems, Foster City, CA). Electrophoresis gel consisted of 12% Long Ranger® (BMA, Rockland, ME), an EDTA-enriched TBE buffer (89 mM Tris, 89 mM borate, and 5.5 mM EDTA), 0.025% ammonium persulfate (APS), and 0.25% N,N,N',N'-tetramethylethylenediamine (TEMED). The remaining conditions were as specified by the manufacturer. Raw data were analyzed using a matrix file for the CUGA dye set and SemiAdaptive basecaller.

$MgCl_2$ (8 mM) can be replaced with $MnCl_2$ (8 mM). The GTP is substituted with 7-deaza-GTP and 1 mM GMP is added as the initiator. TS with T3 RNAP can be done in the same manner. The regions within a span of approximately 50 bp from the cognate promoters cannot be analyzed by using both T7 and T3 RNAPs. The TS protocols with commercially available sequencing kits for both ABI377 and 310 sequencers are provided on the Internet (available at http://www.amebioscience.com/377-EVer1.4.1-EC17-1.pdf and http://www.amebioscience.com/310-EVer2.0-BE22-1.pdf; accessed November 4, 2007).

Overview of Transcriptional Sequencing

The principle of TS is based on Sanger's dideoxy chain termination sequencing method. One difference, though, is the use of RNA polymerase (RNAP) instead of DNAP for the nucleotide incorporation. Another one is the use of 3'-dNTPs, as chain terminators instead of dideoxynucleotides. Hence, sequencers (manual as well as automated) that are applied for the conventional chain termination sequencing can also be used for TS. An initial trial of transcription by using T7 or SP6 RNAP for chain termination sequencing was reported by Axelrod and Kramer (1). They found that the RNA synthesis of a DNA segment, which was inserted immediately downstream from a cognate bacteriophage promoter, by T7 or SP6 RNAP was highly sensitive to the presence of the 3'-dNTP chain terminators. In the course of the investigation on transcription by T7 RNAP, Dröge and Pohl (8) found that the synthetic DNA containing a stretch of $d(CG)_{16}$, which could adopt a left-handed Z-DNA conformation, was efficiently utilized as a template. In this way, they provided evidence that TS could be efficiently applied to sequencing of GC-rich DNA regions.

However, variable incorporation of 3'-dNTPs (dye terminators) caused variations in peak heights and the use of the wild-type RNAP and 3'-dNTPs produced many false ladders due to nonspecific termination. Because of these drawbacks, there has been no further development in the sequencing with the aid of RNAP for over a decade, despite its substantial potential.

Marked advances for accurate and long-read RNAP-based sequencing have been made by introducing newly developed four-color dye-3'-dNTPs (dye terminators) and mutated T7 RNAPs (27). A high incorporation rate of these new dye terminators can be obtained by using the four-color dye-3'-dNTPs spacing with a long carbon linker arm (4- or 6-carbon chains) between nucleotides and fluorescent (rhodamine) dyes. The four-color fluorescent dyes—6-carboxytetramethylrhodamine (TMR), 6-carboxy-X-rhodamine (ROX), 5-carboxyrhodamine-6G (R6G), and 5-carboxyrhodamine-110 (R110)—are used for labeling of 3'-dUTP, 3'-dCTP, 3'-dATP, and 3'-dGTP, respectively (Figure 6-1).

The development of mutated T7 RNAPs has allowed us to improve the uniformity and efficiency of the incorporation rate of dye-3'-dNTPs (see the section on the Characteristics of Modified RNAP). Furthermore, the purified RNase-free yeast pyrophosphatase (PPase) is added into the transcription reaction mixture to inhibit pyrophospholysis that leads to degradation of specific 3'-dNTP-terminated fragments, resulting in an improvement of the peak uniformity (18).

As templates for TS, one can employ ds-DNA fragments that are cloned into a sequencing vector or produced by PCR with primers containing a promoter sequence for T7 or T3 RNAP and the target-specific sequences (Figure 6-2).

Advantages of Transcriptional Sequencing

There are a few obvious advantages of TS for sequencing.

1. *Direct incorporation of the PCR products without purification of sequencing reactions.* In the conventional sequencing system with DNAP, both PCR and sequencing reactions require 2'-dNTPs as substrates. Because a large quantity of unreacted 2'-dNTPs, ddNTPs, and primers used in PCR interfere with the subsequent sequencing reaction, these should be removed by an appropriate procedure such as gel filtration (2). In contrast, TS employs RNAP that incorporates ribonucleoside triphosphates (rNTPs; and finally 3'-dNTPs) instead of 2'-dNTPs, thus enabling transcription initiation by adding the PCR cocktail into the RNAP reaction without any purification of the PCR products.

2. *Short-time isothermal reaction.* In CS, the cycle consisting of heating to denature the strands, annealing of primers and DNA synthesis is repeated, and each of the incubation steps must be carried out at different temperatures. In contrast, the TS reaction is isothermal at 37°C and takes less time compared to the repeated cycling in CS. In addition, RNAP is much more processive (incorporation rate is about 240 bases/sec [22]) compared to Taq polymerase (with

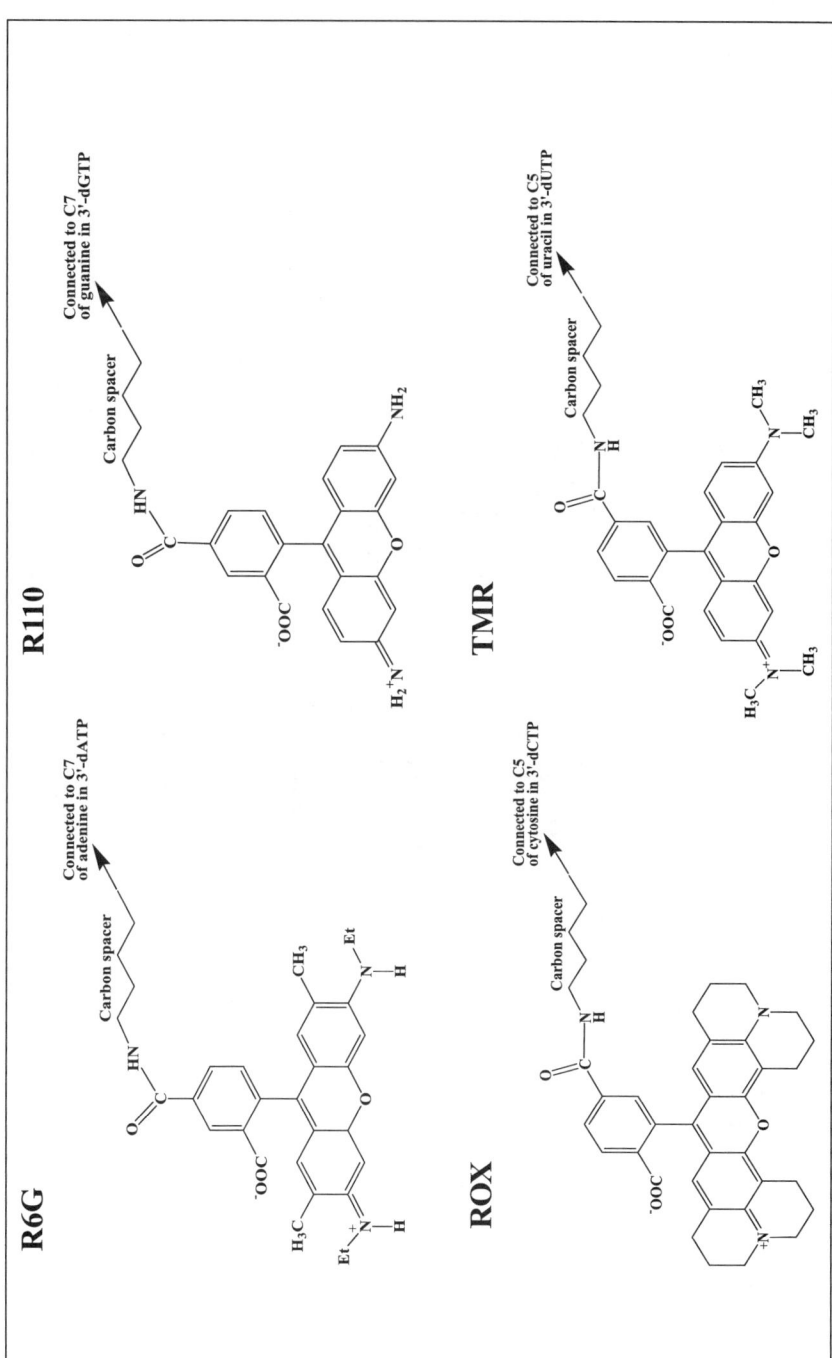

Figure 6-1. **The rhodamine dyes used for the production of transcriptional sequencing terminators.** The rhodamine dyes are attached to dNTPs through a corresponding linker.

Primer design:

5′–CTAATTAACCCTCACTAAAGGG + NNNNNNNNNNNNNNNNNNNNN–3′

       (T3 promoter)   (specific primer)

5′–CTAATACGACTCACTATAGGG + NNNNNNNNNNNNNNNNNNNNN–3′

       (T7 promoter)   (specific primer)

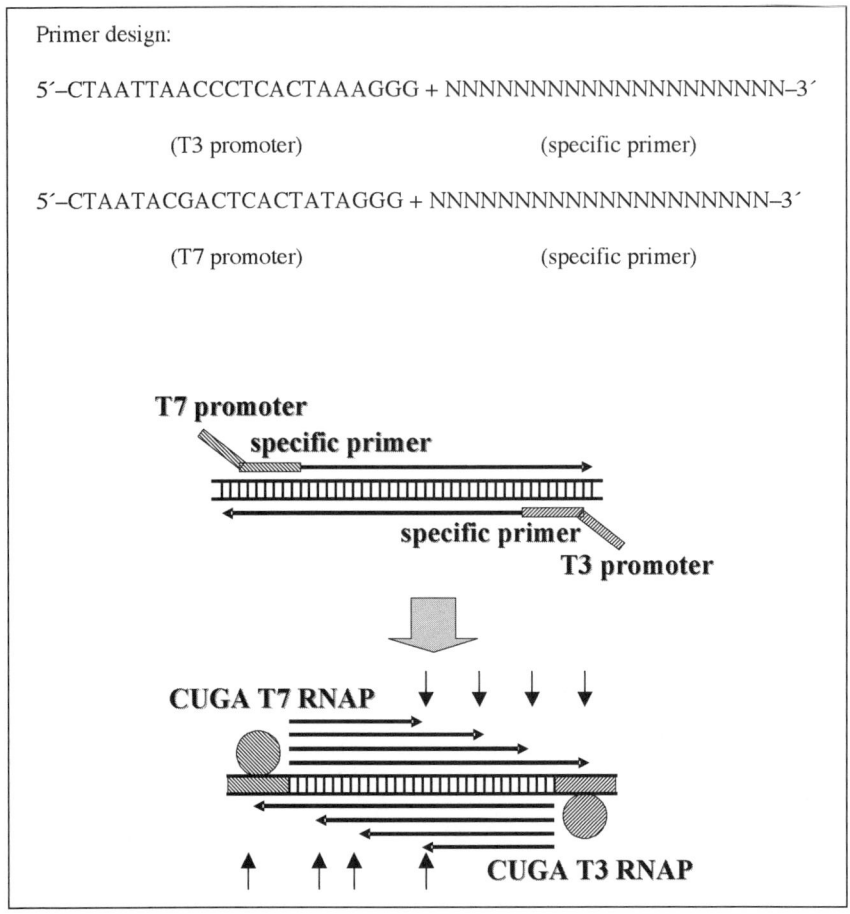

Figure 6-2. Transcriptional sequencing with PCR fragments. The *vertical arrows* indicate the incorporation of 3′-dNTPs as dye-terminators. CUGA is a mutant T7 or T3 RNA polymerase. *Source:* Modified from the manufacturer's brochure.

incorporation rate of about 60 bases/sec [14]), leading to a reduction in reaction time.

3. *High productivity.* A large amount of sequencing products can be transcriptionally amplified from a small amount of DNA templates. Cunningham and Ofengand (5) reported that T7 RNAP can produce 600 molecules of in vitro transcripts from 1 molecule of the DNA template.

4. *Sequencing difficult-to-read regions based on the use of ds-DNA as template.* In the conventional CS method, ds-DNA molecules need to be denatured into single strands for the primer annealing to start the DNA synthesis in the sequencing reaction. When ss-DNA

Table 6-1. Comparison of transcriptional sequencing with conventional cycle sequencing.

	Transcriptional Sequencing	Cycle Sequencing
Enzyme	RNAP	DNAP
Template DNA (in reaction)	Ds-DNA	Ss-DNA
Reaction conditions	Isothermal (37°C)	Non-isothermal (temperature cycling)
Enzymatic recognition sequence	Promoter	Primer binding site
Substrates	rNTPs, 3'-dNTPs	2'-dNTPs, 2',3'-ddNTPs

templates have high GC contents and short repeated sequences, they tend to form higher-order structures during the renaturation, which leads to an interference with the DNA synthesis. Such problems can be greatly alleviated in TS, because ds-DNA can serve as a template and since the RNAPs used in TS can unwind ds-DNA templates during the transcription reaction. Table 6-1 summarizes the comparison between TS and CS.

Characteristics of Modified RNAP

One of the major breakthroughs in sequencing through transcription is the development of mutant or modified T7 RNAPs that are capable of a highly accurate and efficient synthesis of RNA. Amino acid residues (phenylalanines at positions 644 and 667) responsible for the high selectivity of T7 RNAP for rNTPs have been identified through the construction of various mutants and by evaluating their capability to synthesize RNA (18). These sites interact with the 3'-OH group of rNTP in T7 RNAP on the basis of three-dimensional crystal structure analysis (32). The mutant analysis has shown that the F644Y and F667Y substitutions bring more than 200-fold higher affinities compared to the wild type for 3'-dNTPs incorporation (18, 27). The mutant T7 RNAP, included in the commercial TS kit, has been developed through further modifications. Modified T3 RNAPs have also been prepared by substitution of phenylalanine at 645 and 668 with tyrosine and deletion of lysine 173 and/or arginine 174 or one of the amino acids located at residues 179–181.

The modified T7 RNAP has several features favorable to *in vitro* transcription.

1. It can synthesize specific RNA molecules in a large quantity: more than 200 µg of RNA molecules with a linear DNA template at 37°C in 2 hours.
2. It can synthesize authentic or exact copies of long and short RNA molecules. The amount of aberrant transcription products with nonspecifically elongated 3'-termini is reduced significantly compared to similar transcripts obtained with the wild type T7 RNAP.
3. An efficient transcription by the modified T7 RNAP is independent of the terminal structures of the templates. In the conventional *in vitro* RNA synthesis, production of large extraneous RNA species is often observed when DNAs having a protruding 3'-terminus are used as templates (28). In contrast, the *in vitro* transcription system that uses the T7 RNAP mutant results in an accurate RNA synthesis independently of the terminal structures of the DNA templates.
4. It can elongate independently of the template sequences and the recognition of transcriptional terminators by the T7 RNAP mutant is much stricter than that by the wild type RNAP, which leads to production of exact copies of the RNA molecule in question.

In vitro transcription tends to stop where a terminator or some like sequence is located. However, the band compression can be minimized in the TS system as a result of the use of inosine 5'-triphosphate instead of guanosine 5'-triphosphate as the substrate to interfere with the formation of secondary structures that may trigger transcription termination in the nascent RNA. TS reaction could still be further improved. For example, a series of ethylated polyamine analogues including 1,8-bis(ethylamino)octane and 1,8-octanediamine work as active enhancers in TS (16). It has been suggested that the T7 RNA polymerase could be activated by the specific binding of the polyamine additive to produce RNA transcripts with high fidelity to the template DNA.

Sequencing of Difficult-to-Read Templates by the Transcriptional Sequencing Method

The effectiveness of TS in the sequencing of difficult-to-read templates can be demonstrated with several criteria. The most successful examples can be seen when TS is applied to the sequencing of GC-rich stretches. When various mouse cDNA clones whose chromatogram peaks suddenly dropped off (Figure 6-3, top) or became weak (Figure 6-4, top) in the conventional CS were subjected to TS, all of the sequences were accurately

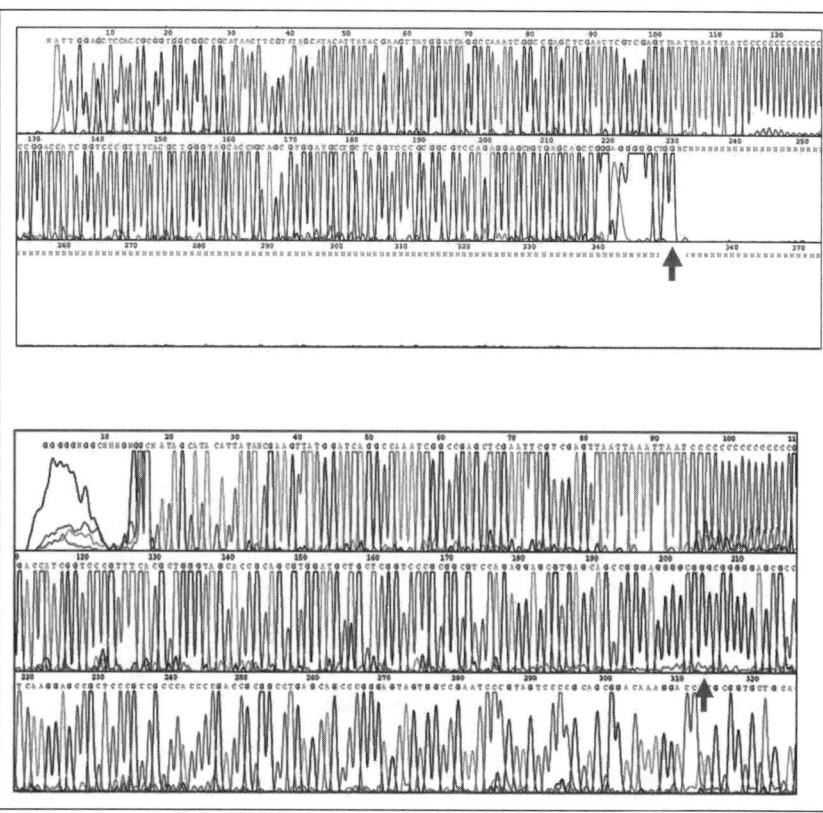

Figure 6-3. **Sudden signal drop off pattern (see Plate 9 in the Color Addendum).** Top: Reading suddenly stops just after a GC-rich stretch (*arrow*) in CS. Bottom: Overcoming the GC-rich stretch (*arrow*) by TS. *Source:* From Shibata, K., Izawa, M., Hayashizaki, Y., et al. 2003. Practical application of transcriptional sequencing for GC-rich templates. *J Struct Funct Genomics* 4: 35–39. Reprinted with kind permission of Springer Science and Business Media.

determined without pausing (Figures 6-3 and 6-4, bottom), even if the GC content of each of the regions ranging from +20 to −20 (where the termination points in CS is +1) exceeded 85% (30).

Ishikawa and colleagues tested TS with three kinds of synthetic oligonucleotide templates with two or more of the following structural features: high GC content (80%), a homopolymeric sequence of G and C, a dual repeated GC cluster sequence, or a putative hairpin (inverted repeat) structure of 30 bp (15). Complete sequences were obtained with all of these troublesome oligonucleotide templates by TS. With BigDye terminator chemistry, sequencing signals in all templates were significantly reduced and suddenly became unreadable, for example, just after encountering a

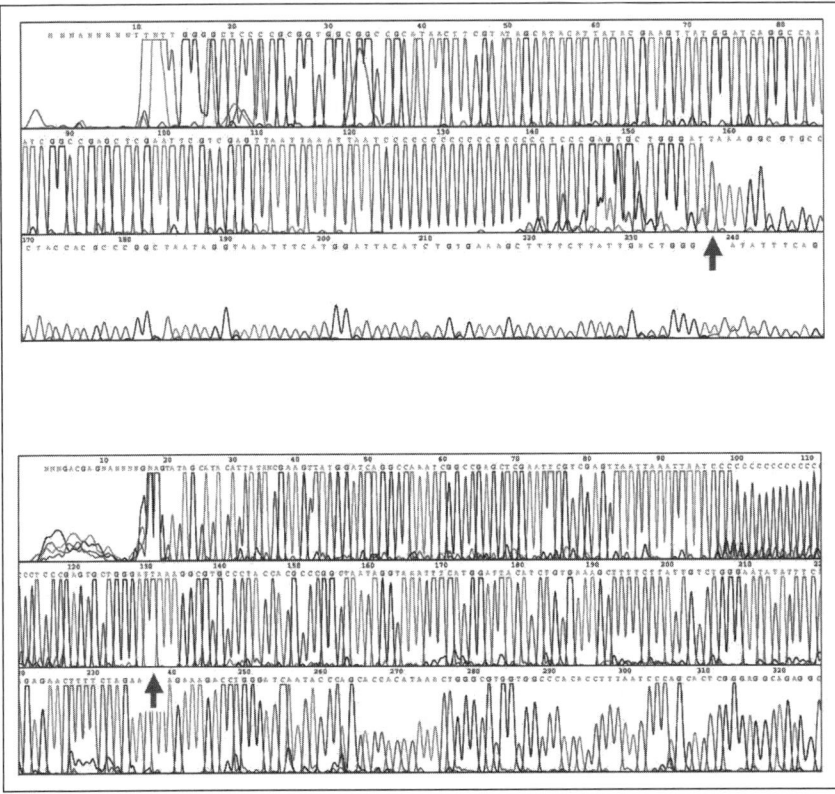

Figure 6-4. Sudden signal weakening pattern (see Plate 10 in the Color Addendum). Top: Signals suddenly weaken just after a GC-rich stretch (*arrow*) in CS.
Bottom: Overcoming the GC-rich stretch (*arrow*) by TS. *Source:* From Shibata, K.,
Izawa, M., Hayashizaki, Y., et al. 2003. Practical application of transcriptional
sequencing for GC-rich templates. *J Struct Funct Genomics* 4: 35–39. Reprinted with
kind permission of Springer Science and Business Media.

homopolymeric G region. TS also can be successfully applied to a cDNA
clone containing GC contents as high as 88% in the part of the 5′-
untranslated region and 74% in whole cDNA. A template containing an
inverted repeat sequence, in which a hairpin structure with a 120-base
stem and an approximately 300-base loop was predicted, was also perfectly resolved by TS (15). Moreover, TS was effective for a simple sequence
repeat containing $(TG)_{27}$ and $(AG)_{28}$ continuously. Recently, it has been
reported that TS was successfully employed to sequence DNA templates
of the *Euglena gracilis* fibrillarin gene that were particularly difficult to
read accurately using conventional sequencing protocols (26). In addition,
the recent manufacturer's brochure reported that the 126-bp sequence of

GT-repeats can be resolved by TS (available at http://www.nippongene-tech.com; accessed November 4, 2007).

The significant inapplicability of CS chemistry to sequence these difficult-to-read regions is thought to be due mainly to the denaturation leading to the formation of stable hairpin structures of the ss-DNA template. DNAP should stop at such strong secondary structures that are followed by pausing, terminating strand extension or skipping. It is sometimes hard to resolve long A- or T-stretches such as AAAAAAAAA ... or TTTTTTTT ... even by using TS. This failure is due to slippage characteristics of the transcription reaction by RNAP (21) and a major problem to be resolved in TS. It has been suggested that the template strand sequence determines the polymerase slippage (20).

Multiple Displacement Amplification (MDA)-TS

An improved protocol permitting rapid TS of highly GC-rich regions with support of the Multiple Displacement Amplification (MDA) has recently been reported (17). MDA exponentially amplifies circular DNA templates through the random-primed rolling circle mechanism with bacteriophage phi29 DNA polymerase (6, 7) and is an alternative to general plasmid preparation. Using random primers, circular DNA templates can be amplified 10,000-fold in a few hours. PCR on MDA-amplified DNA was successfully done for challenging long DNA segments (>10 kb) with high GC content (~80%) (34). A five-minute TS reaction with the MDA-amplified human epidermal growth factor receptor gene was sufficient to get 100% sequence data of its deleted regions associated with Iressa (Gefitinib)-sensitivity in non-small cell lung carcinoma (17).

MDA also favorably compares with PCR in the amplification of various human GC-rich templates. The former was found to be superior in the success rate and specific band formation (17). It should be noted that MDA-PCR could amplify the templates that gave only unsatisfactory results even by using extended PCR cycles and addition of several GC-destabilizing agents, including DMSO.

A new plasmid vector convenient for MDA-TS has been constructed. The vector, pTS1 (Figure 6-5), which was derived from pUC19 and carries T7 as well as T3 promoters, contains the regions from −17 to +6 of the promoter sequences as the minimal region essential for strong transcriptional activity. In most of the commercially available vectors such as pBS (Stratagene, La Jolla, CA) and pGEM (Promega, Wisconsin), the T7 promoter sequences in the +4 to +6 regions are different from the consensus promoter sequences and sequence variations in the +1 to +6 regions are known to decrease the transcription efficiency (13). The multi-cloning site is flanked by each of the bacteriophage promoters in order to sequence both strands of DNA template. A comparison between these T7

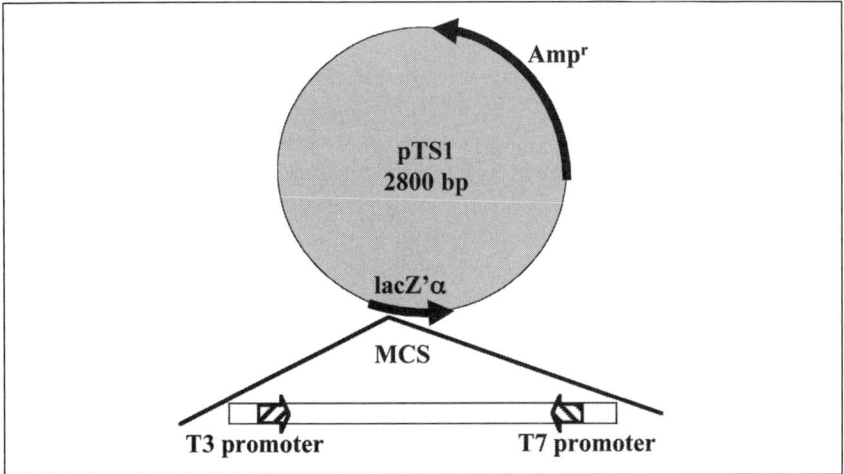

Figure 6-5. **The pTS1 vector for bidirectional TS sequencing.** There are recognition sequences for various restriction endonucleases in the multicloning sites (MCS): *Bss*HII, *Sac*I, *Pst*I, *Nhe*I, *Bam*HI, *Stu*I, *Hin*dIII, *Hinc*II, *Eco*RI, *Sma*I, *Bgl*III, *Sna*BI, *MLu*I, *Kpn*I, *Nsi*I, *Bss*HII in the 5′-to-3′ orientation. The *Bss*HII sites are located on the outside of the T3 and T7 promoters.

promoter-containing vectors revealed the highest accuracy (>99.5%) in sequencing with pTS1. TS sequencing with pTS1 plasmids containing a GC-rich or a GT repetitive region gave a clean sequence pattern without a decrease in signal at the specific region where signal peaks suddenly disappeared when using CS (17).

Conclusions

Sequencing projects of a large number of organismic genomes have been completed, but there are many gaps yet to be resolved in most of them, even in the human genome (35). The difficult-to-read sequences using CS are mostly concentrated in regions with high GC content or in highly ordered secondary structures. Some of the genomic regions associated with transcriptional regulation correspond to *cis*-regulatory GC-rich regions such as CpG islands and promoters themselves (10). In view of the fact that around 72% of the mouse genome regions can produce the corresponding transcripts (4), at least a part of the unresolved unsequenced genomic regions may be involved in various regulatory mechanisms in the nucleus. Because of its superior resolving power, TS could be a rational choice for examining if these difficult-to-sequence stretches have important biological information.

The TS kit maker is now developing kits that will be adaptable for TS with sequencers other than ABI 370XL and 310. We expect that the TS system would be optimized for use with modern sequencing machines such as ABI3100, 3730, and 3730XL in the near future. This will increase the value of TS in resolving a number of difficult-to-read sequences of a wide variety of organisms.

Acknowledgments

The author thanks Kazuhiro Shibata and Masaki Izawa for comments. This work was supported by grants for the RIKEN Genome Exploration Research Project and Genome Network Project.

References

1. Axelrod, V.D., and Kramer, F.R. 1985. Transcription from bacteriophage T7 and SP6 RNA polymerase promoters in the presence of 3′-deoxyribonucleoside 5′-triphosphate chain terminators. *Biochemistry* 24: 5716–5723.
2. Bachmann, B., Lüke, W., and Hunsmann, G. 1990. Improvement of PCR amplified DNA sequencing with the aid of detergents. *Nucleic Acids Res* 18: 1309.
3. Baskaran, N., Kandpal, R.P., Bhargava, A.K., et al. 1996. Uniform amplification of a mixture of deoxyribonucleic acids with varying GC content. *Genome Res* 6: 633–638.
4. Carninci, P., Kasukawa, T., Katayama, S., et al. 2005. The transcriptional landscape of the mammalian genome. *Science* 309: 1559–1563.
5. Cunningham, P.R., and Ofengand, J. 1990. Use of inorganic pyrophosphatase to improve the yield of in vitro transcription reactions catalyzed by T7 RNA polymerase. *Biotechniques* 9: 713–714.
6. Dean, F.B., Hosono, S., Fang, L., et al. 2002. Comprehensive human genome amplification using multiple displacement amplification. *Proc Natl Acad Sci U S A* 99: 5261–5266.
7. Dean, F.B., Nelson, J.R., Giesler, T.L., et al. 2001. Rapid amplification of plasmid and phage DNA using Phi 29 DNA polymerase and multiply-primed rolling circle amplification. *Genome Res* 11: 1095–1099.
8. Dröge, P., and Pohl, F.M. 1991. The influence of an alternate template conformation on elongating phage T7 RNA polymerase. *Nucleic Acids Res* 19: 5301–5306.
9. Fernandez-Rachubinski, F., Eng, B., Murray, W.W., et al. 1990. Incorporation of 7-deaza dGTP during the amplification step in the polymerase chain reaction procedure improves subsequent DNA sequencing. *DNA Seq* 1: 137–140.
10. Hapgood, J.P., Riedemann, J., and Scherer, S.D. 2001. Regulation of gene expression by GC-rich DNA cis-elements. *Cell Biol Int* 25: 17–31.

11. Haqqi, T., Zhao, X., Panciu, A., et al. 2002. Sequencing in the presence of betaine: Improvement in sequencing of the localized repeat sequence regions. *J Biomol Tech* 13: 265–271.

12. Henke, W., Herdel, K., Jung, K., et al. 1997. Betaine improves the PCR amplification of GC-rich DNA sequences. *Nucleic Acids Res* 25: 3957–3958.

13. Imburgio, D., Rong, M., Ma, K., et al. 2000. Studies of promoter recognition and start site selection by T7 RNA polymerase using a comprehensive collection of promoter variants. *Biochemistry* 39: 10419–10430.

14. Innis, M.A., Myambo, K.B., Gelfand, D.H., et al. 1988. DNA sequencing with *Thermus aquaticus* DNA polymerase and direct sequencing of polymerase chain reaction-amplified DNA. *Proc Natl Acad Sci U S A* 85: 9436–9440.

15. Ishikawa, T., Hayashida, Y., Hirayasu, K., et al. 2003. Use of transcriptional sequencing in difficult to read areas of the genome. *Anal Biochem* 316: 202–207.

16. Iwata, M., Izawa, M., Sasaki, N., et al. 2000. T7 RNA polymerase activation and improvement of the transcriptional sequencing by polyamines. *Bioorg Med Chem* 8: 2185–2194.

17. Izawa, M., Kitamura, N., Odake, N., et al. 2006. A rapid and simple transcriptional sequencing method for GC-rich DNA regions. *Jpn J Vet Res* 53: 159–168.

18. Izawa, M., Sasaki, N., Watahiki, M., et al. 1998. Recognition sites of 3'-OH group by T7 RNA polymerase and its application to transcriptional sequencing. *J Biol Chem* 273: 14242–14246.

19. Kieleczawa, J. 2006. Fundamentals of sequencing of difficult templates—an overview. *J Biomol Tech* 17: 207–217.

20. Kwon, Y.S., and Kang, C. 1999. Bipartite modular structure of intrinsic, RNA hairpin-independent termination signal for phage RNA polymerases. *J Biol Chem* 274: 29149–29155.

21. Macdonald, L.E., Zhou, Y., and McAllister, W.T. 1993. Termination and slippage by bacteriophage T7 RNA polymerase. *J Mol Biol* 232: 1030–1047.

22. Makarova, O.V., Makarov, E.M., Sousa, R., et al. 1995. Transcribing of *Escherichia coli* genes with mutant T7 RNA polymerases: stability of *lacZ* mRNA inversely correlates with polymerase speed. *Proc Natl Acad Sci U S A* 92: 12250–12254.

23. Mizusawa, S., Nishimura, S., and Seela, F. 1986. Improvement of the dideoxy chain termination method of DNA sequencing by use of deoxy-7-deazaguanosine triphosphate in place of dGTP. *Nucleic Acids Res* 14: 1319–1324.

24. Mytelka, D.S., and Chamberlin, M.J. 1996. Analysis and suppression of DNA polymerase pauses associated with a trinucleotide consensus. *Nucleic Acids Res* 24: 2774–2781.

25. Potaman, V.N., and Bissler, J.J. 1999. Overcoming a barrier for DNA polymerization in triplex-forming sequences. *Nucleic Acids Res* 27: e5.

26. Russell, A.G., Watanabe, Y., Charette, J.M., et al. 2005. Unusual features of fibrillarin cDNA and gene structure in *Euglena gracilis*: evolutionary conservation of core proteins and structural predictions for methylation-guide box C/D snoRNPs throughout the domain Eucarya. *Nucleic Acids Res* 33: 2781–2791.

27. Sasaki, N., Izawa, M., Watahiki, M., et al. 1998. Transcriptional sequencing: a method for DNA sequencing using RNA polymerase. *Proc Natl Acad Sci U S A* 95: 3455–3460.

28. Schenborn, E.T., and Mierendorf, R.C. Jr. 1985. A novel transcription property of SP6 and T7 RNA polymerases: dependence on template structure. *Nucleic Acids Res* 13: 6223–6236.

29. Seto, D., Seto, J., Deshpande, P., et al. 1995. DMSO resolves certain compressions and signal dropouts in fluorescent dye labeled primer-based DNA sequencing reactions. *DNA Seq* 5: 131–140.

30. Shibata, K., Izawa, M., Hayashizaki, Y., et al. 2003. Practical application of transcriptional sequencing for GC-rich templates. *J Struct Funct Genomics* 4: 35–39.

31. Slesarev, A.I., Mezhevaya, K.V., Makarova, K.S., et al. 2002. The complete genome of hyperthermophile *Methanopyrus kandleri* AV19 and monophyly of archaeal methanogens. *Proc Natl Acad Sci U S A* 99: 4644–4649.

32. Sousa, R., Chung, Y.J., Rose, J.P., et al. 1993. Crystal structure of bacteriophage T7 RNA polymerase at 3.3 A resolution. *Nature* 364: 593–599.

33. Turner, S.L., and Jenkins, F.J. 1995. Use of deoxyinosine in PCR to improve amplification of GC-rich DNA. *Biotechniques* 19: 48–52.

34. Yan, J., Feng, J., Hosono, S., et al. 2004. Assessment of multiple displacement amplification in molecular epidemiology. *Biotechniques* 37: 136–143.

35. Zody, M.C., Garber, M., Sharpe, T., et al. 2006. Analysis of the DNA sequence and duplication history of human chromosome 15. *Nature* 440: 671–675.

Bias-Free Cloning of "Unclonable" DNA for Simplified Genomic Finishing

Ronald Godiska,[1] David Mead,[1]
Vinay Dhodda,[1] Rebecca Hochstein,[1]
Attila Karsi,[2] Nikolai Ravin,[3]
and Chengcang Wu[1]
*[1]Lucigen Corp., Middleton, WI, and [2]Mississippi
State University, Mississippi State, MS, and
[3]Center Bioengineering, Russian Academy of
Science, Moscow, Russia*

A major component of genomic sequencing is the finishing process, which entails filling in sequence gaps that remain after high-throughput shotgun cloning and sequencing. After sequencing to 8× to 10× coverage, a typical microbial genome of 2 Mb may have more than 100 gaps, and a mammalian sequence of 3 Gb may have >150,000 gaps (20, 42). Gaps due to difficulties in sequencing have been addressed elsewhere in this volume. However, another major source of gaps is the inherent bias of current cloning processes, as several classes of DNA sequences are unstable or unclonable in conventional plasmid vectors (12). In some cases, genomic regions encompassing millions of bases are excluded from the libraries (15). Even after the finishing process seems to be complete, there may be hundreds or thousands of gaps remaining in the sequence. For example, the "finished" draft of the human genome contained >400 gaps (14, 15, 39). Consequently, obtaining the sequence of the final 5% of a genome often requires an inordinate amount of effort (4).

Numerous types of genetic elements are subject to this instability. In fact, nearly the entire genome may be difficult to clone for some organisms, such as the malaria parasite *Plasmodium* (10), *Bacillus subtilis* (19), and *Dictyostelium discoideum* (11). Common features of difficult templates

DNA III: Dealing with Difficult Templates
Edited by Jan Kieleczawa
©2008 Jones and Bartlett Publishers

include high AT-content, strong secondary structure, repetitive DNA, or *cis*-acting functions (e.g., transcriptional promoters or replication origins). Segments containing open reading frames can be difficult to clone as well, if fortuitous expression of the encoded peptide is detrimental to the host cell.

The inability to clone these templates is often exacerbated by features inherent to conventional cloning vectors, such as pUC19 and its derivatives (12). The "blue/white" colony screen drives strong expression of cloned sequences, selecting against many coding regions. It also destabilizes a variety of secondary structures, such as di- or trinucleotide repeats and poly-T tracts (2, 3, 16, 17, 23, 28, 30). Cloned promoters can drive transcription into the vector backbone, inhibiting expression of the selectable marker gene (20) or impeding the function of the plasmid replication origin (40). The high copy number of conventional vectors is detrimental for cloning fragments that are large (e.g., >8 kb) or capable of forming secondary structure (5, 9; data not shown). The high copy number also may promote recombination among plasmids. Torsional strain caused by supercoiling can induce formation of cruciforms and other secondary structures within tandem repeats and palindromic sequences, resulting in deletions and instability (21, 24).

Large-insert bacterial artificial chromosomes (BACs) circumvent some of these problems. Because of the large size of the inserts (>200 kb), these libraries have been essential in finishing the genomic sequence of all large and complex genomes. They are the basis of whole genome physical mapping, clone-by-clone sequencing, and map assembly in the whole genome shotgun approach (26, 44–48). Currently, all BAC libraries rely on partial restriction digestion to fragment the DNA. However, the restriction sites are biased in the genomes of all microbial, plant, and animal species studied. Indeed, tracts of highly repetitive regions, such as centromeres and telomeres, may completely lack such sites. These regions account for clone gaps that can span megabases of genomic DNA. Regions with extremely biased GC content, high methylation, and other modifications may also be under- or over-digested. Biased representation of genomic regions is a pervasive problem of conventional BAC libraries.

We demonstrate here several novel approaches to alleviate these difficulties, allowing much more efficient genome assembly. A novel linear cloning vector, the "pJAZZ" plasmid, routinely maintains fragments that appear impossible to clone in conventional supercoiled vectors. This vector efficiently clones inserts up to 30 kb, allowing assembly of microbial genomes with minimal finishing and without the need for fosmids or BAC libraries. A new method of BAC library construction, based upon random shearing of genomic DNA, permits capture of highly repetitive DNA, such as centromeric regions. Random Shear BAC cloning has been used to close several centromeric gaps in the "finished" sequence of the

Arabidopsis genome. In addition to these improvements, transcription-free cloning vectors provide stable cloning of several types of targets that are otherwise deleted.

Materials and Methods

Transcription-Free Cloning with the Linear pJAZZ Vector

Host Strain

The *E. coli* strain BigEasy TSA is the host for the pJAZZ vectors. These cells were derived from DH10B cells by chromosomal integration of three loci from phage N15: the protelomerase gene *tel*N, the partition genes *sopAB*, and the *ant*A anti-repressor gene, the latter under control of the *ara*BAD promoter. These cells provide stable, inducible replication of the pJAZZ vectors upon growth in 0.01% arabinose.

Cloning

The pJAZZ vector was prepared for cloning by digestion with *Sma*I (blunt) or *Ahd*I (3′C overhang), which excises the *lacZ* stuffer from the vector arms. The digests were dephosphorylated with Calf Intestinal Phosphatase (Promega, Madison, WI) and purified by binding to the Qiaquick PCR Cleanup column (Qiagen, Valencia, CA). For ligation, 50 to 100 ng of digested linear vector and 100 to 300 ng of insert DNA were ligated with the CloneDirect ligation kit (Lucigen, Middleton, WI).

pJAZZ Library Construction

Genomic DNA was sheared to the desired size range with a HydroShear device (Genomic Solutions, Ann Arbor, MI). The sheared DNA was end-repaired with the DNATerminator Kit (Lucigen) and fractionated by electrophoresis on a 1% agarose gel. Fractions were excised and purified (Gel Extraction Kit; Qiagen), and ligated to a *Sma*I digest of the pJAZZ vector. For some libraries, the blunt DNA was treated with PyroPhage™ DNA polymerase (Lucigen) in the presence of dGTP to add single 3′G tails to the fragments. The G-tailed DNA was ligated to an *Ahd*I digest of the pJAZZ vector. Ligations were electroporated into BigEasy TSA cells. Recombinants in pJAZZ-OC were selected on plates containing 12.5 μg/mL chloramphenicol plus X-GAL (Amresco, Solon, OH); pJAZZ-KA recombinants were selected on plates containing 100 μg/mL ampicillin, 20 μg/mL kanamycin, and X-GAL.

The pJAZZ clones typically produced large colonies and grew vigorously in culture. After induction of copy number, clones yielded 5 to 20 μg of linear plasmid DNA from standard 1.5-mL alkaline lysis minipreps. For gel analysis, one-tenth to one-fifth of the DNA from each miniprep was digested with NotI to excise the insert. Inserts generally reflected the size range of the input DNA. Sequencing reactions were performed with dye terminator chemistry according to the manufacturer's instructions (Amersham, Piscataway, NJ), using 150 to 250 ng of template DNA from linear vector clones.

Results

Vector Construction

Phage N15 is a 46-kb double-stranded DNA phage with cohesive ends (38). Its head and tail genes have extensive homology with those of bacteriophage lambda, as do the elements for control of transcription and prophage immunity (37). The N15 genome contains a partition system (*sopAB*), which has homology to F' plasmid SOP genes, but its centromere sites are dispersed throughout the phage genome (8, 31).

The replication mechanism of N15 is unique (32, 36). N15 prophage replicates as a linear dsDNA molecule with covalently closed hairpin ends (Figure 7-1) (36). The only N15 gene required for replication is *rep*A, which contains helicase, primase, and origin-binding activities (25). The origin itself is within the *rep*A gene. Replication proceeds bidirectionally using the host *E. coli* DNA polymerase (41). After replication of the ends, the TelN pro-telomerase of N15 cleaves at the palindromic site created at each of the telomeres. The 5' and 3' strands of each newly formed end are joined by TelN, reforming a closed hairpin telomere (see Figure 7-1) (6, 35).

The linear vector pG591, previously described by Ravin et al. (33, 34), contains the genes *tel*N, *rep*A, and *cB* (copy number regulator) of phage N15, but it lacks the phage structural genes. Although pG591 is functional as a vector, the left arm alone is capable of transforming cells, generating an unacceptable background of dimers or circular permutations (data not shown). The cloning site of pG591 also allows transcriptional interference between the vector and the insert, which may substantially detract from stability of inserts (12).

The linear vectors pJAZZ-KA and pJAZZ-OC were created to increase the utility of pG591 (see Figure 7-1). Numerous improvements were incorporated, including: (a) transcriptional terminators to protect the vector from insert-driven transcription; (b) a terminator to block transcription into the insert from a vector promoter; (c) a small multiple cloning site; (d) drug selection for the right arm of the vector; and (e) a *lacZ* "stuffer" fragment, which is excised from the vector preparation, to screen against uncut vector. To decrease the size of the vector, the drug resistance marker

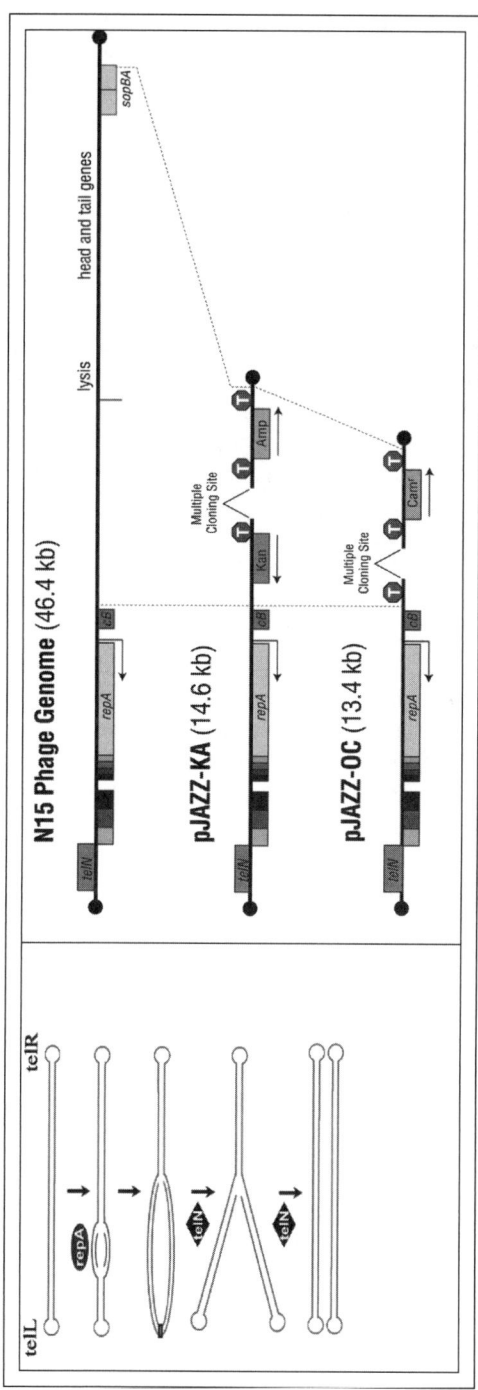

Figure 7-1. Replication and structure of phage N15 and linear pJAZZ vectors. Left Panel: Replication of N15 and the pJAZZ vectors requires the phage proteins RepA and TelN. Right Panel: The lysis and structural genes of N15 were replaced with selectable markers and a cloning site. A *lacZ* "stuffer" fragment, situated between a pair of identical multiple cloning sites, is removed before ligation to target DNAs. Transcription is indicated by arrows and transcriptional terminators by "T." Dark balls represent the hairpin telomeres.

was removed from the left arm of the vector in pJAZZ-OC. The origin of replication is essential for viability, so it serves as a selectable marker for the left arm. Except for the difference in drug selection, the two versions of pJAZZ appear to behave identically in cloning applications (data not shown).

Cloning Large AT-Rich DNA

The AT-rich genome of *Lactobacillus helveticus* (67% AT) has been very difficult to clone in conventional vectors. Inserts as small as 1 to 3 kb were unstable in pUC19, often yielding clones that were smaller than the parental vector (Figure 7-2a). To determine whether transcription affected the stability of this DNA, the fragments were cloned into the transcription-free, circular vector pSMART-HCKan (Figure 7-2b). Inserts of up to approximately 1 to 3 kb were stable in this vector (Figure 7-2c). Additional results showed that clones of 4 to 6 kb were successfully maintained only in the low-copy version of the pSMART vector (data not shown). Markedly improved results were obtained with the linear pJAZZ-KA vector, as stable inserts of >20 kb were routinely recovered (Figure 7-2d).

Similar results were obtained with genomic DNA from the ciliated protozoans *Tetrahymena* and *Oxytricha*. The AT content of these genomes is 75% to 85%, and fragments of >6 kb have been extremely problematic to clone in traditional circular plasmid vectors (7). Even in the best case, attempts to clone this DNA into a transcription-free BAC vector resulted in very few recombinants, most of which grew poorly and were unstable for long-term propagation (data not shown). To test whether this DNA could be stably cloned into a linear plasmid, the vector pJAZZ-KA was used to construct large-insert libraries of randomly sheared DNA from these two genomes. Obtaining stable inserts in the range of 6 to 20 kb presented no apparent difficulty with this vector (Figure 7-3a). The clones

---

Figure 7-2. Increased stability of AT-rich DNA in transcription-free and linear vectors. (a) Genomic DNA from *L. helveticus* was sheared to 1 to 3 kb and cloned into pUC19. Transformants were frequently deleted, yielding clones smaller than the empty vector control (lane "V"). Uncut DNAs are shown. (b) The circular vectors pSMART-HCKan (1.8 kb) or pSMART-LCKan (2.1 kb) differ only by the presence of the *ROP* gene (repressor of primer) in the low copy version of the vector. (c) Random shotgun inserts 1 to 3 kb of *L. helveticus* DNA were stable in the pSMART-HCKan vector, as all clones were 1 to 3 kb larger than the empty vector. Uncut DNAs are shown. (d) Genomic DNA from *L. helveticus* was sheared to 10 to 20 kb and cloned into the pJAZZ vector. Clones were digested with *Not*I to excise the inserts. Left arm of linear vector is 12 kb (*); the right arm ran off the gel. M, size markers.

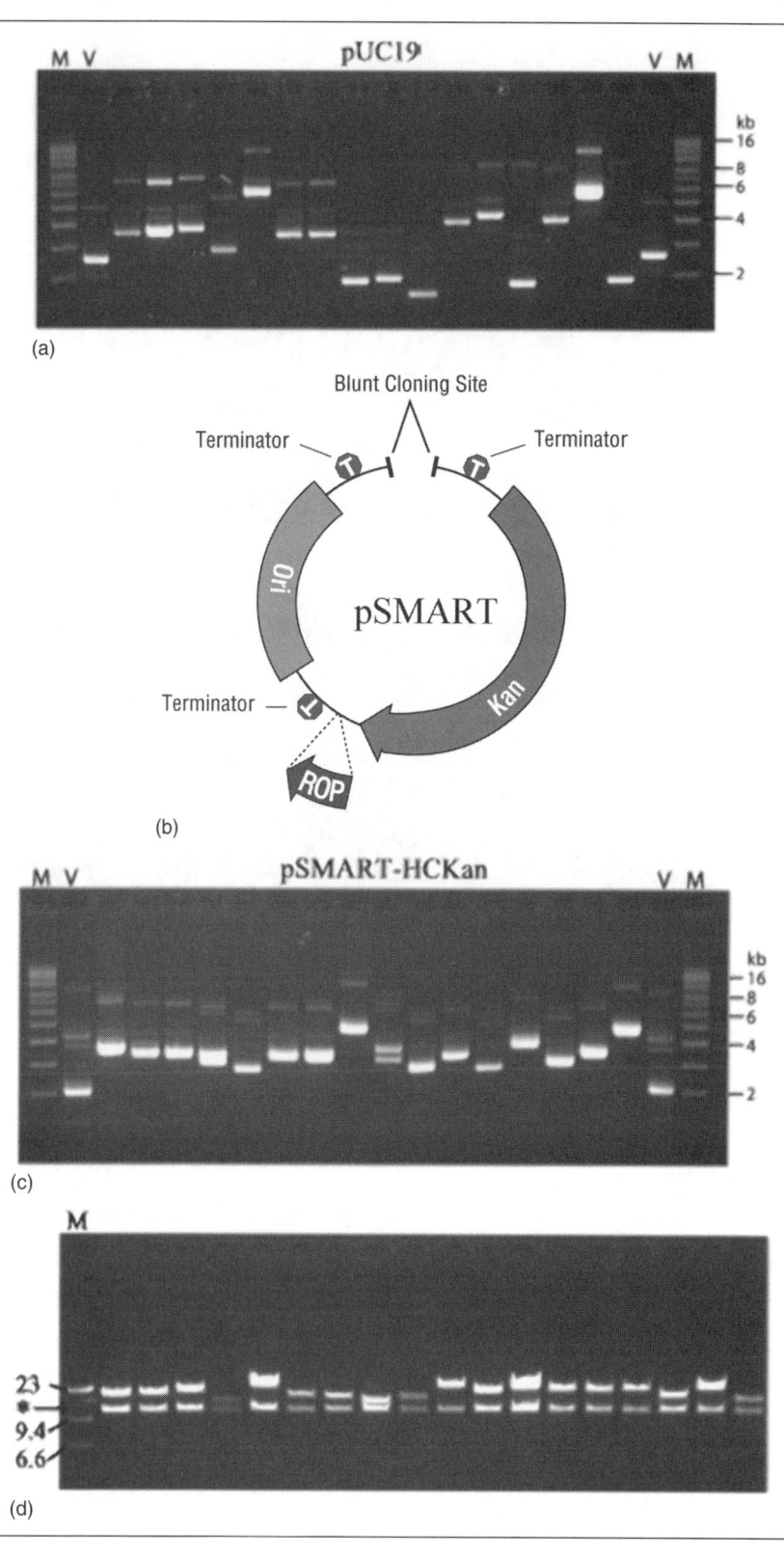

(a)

pUC19

kb
16
8
6
4
2

Blunt Cloning Site

Terminator   Terminator

Ori   **pSMART**   Kan

Terminator — T

ROP

(b)

pSMART-HCKan

kb
16
8
6
4
2

(c)

23
*
9.4
6.6

(d)

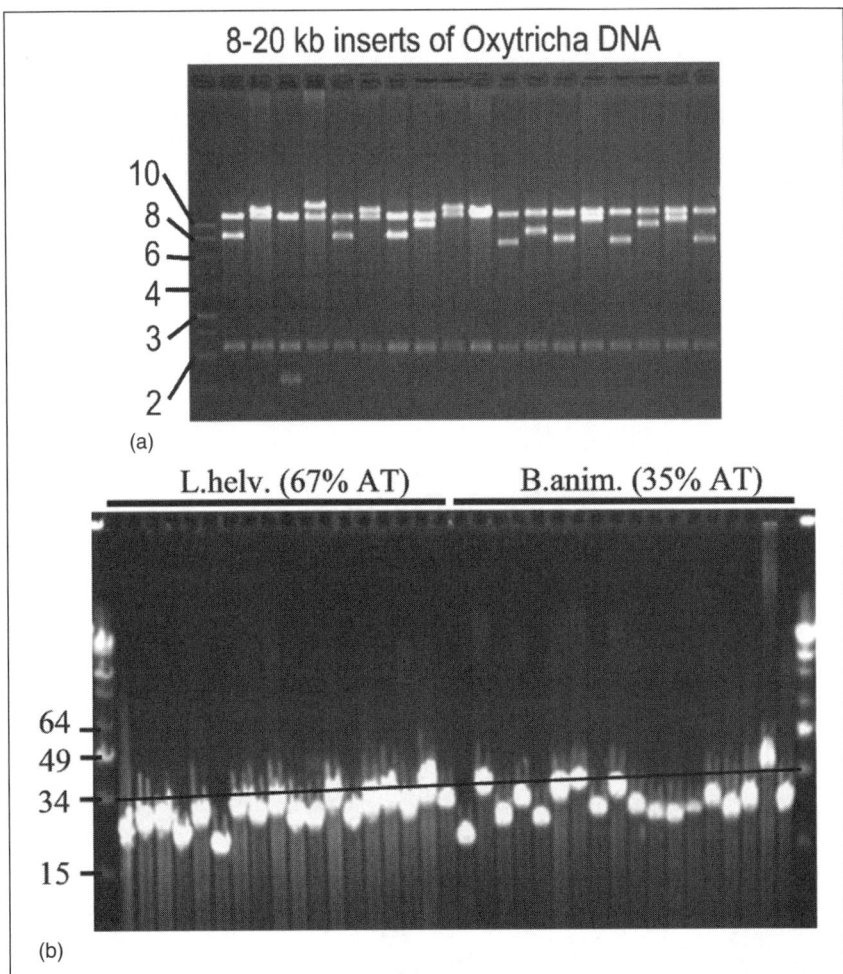

Figure 7-3. **Large AT-rich or GC-rich inserts in the pJAZZ vector.** Genomic DNA was sheared, end-repaired, and cloned into the pJAZZ vector. (a) *Oxytricha* library, showing that most inserts have a size of 6 to 20 kb. Each lane contains 1/5th of the DNA from a 1.5-mL miniprep, cut with *Not*I to excise the insert. Vector bands are 12 kb and 2.2 kb. (b) Libraries from *L. helveticus* or *B. animalis*. Uncut DNA from recombinants migrated at 22 to 34 kb, corresponding to inserts of 8 to 20 kb.

produced large colonies, grew vigorously in culture, and yielded relatively high amounts of linear plasmid DNA. To our knowledge, the linear pJAZZ libraries represent the first successful large-insert libraries of *Oxytricha* or *Tetrahymena* DNA, despite numerous attempts by us and by others using traditional cloning systems.

Cloning Large GC-Rich DNA

Although problematic genomes often tend to be AT-rich, some GC-rich genomes have presented difficulties as well. An example of a genome with relatively high GC content is *Bifidobacterium animalis* (65% GC). Libraries created in the pJAZZ vector produced stable, large inserts (Figure 7-3b), confirming that the vector is suitable for cloning GC-rich fragments.

Repetitive DNA

Another class of inserts that is extremely difficult to clone in circular plasmids is highly repetitive DNA. An example is a cDNA library that we constructed from snail tissue. The cDNA was size-selected to 0.3–0.7 kb and 0.7–2.0 kb. The fractions initially were cloned into a pUC19 derivative or into the pSMART-HCKan or –LCKan circular vectors (high or low copy number, respectively). All inserts recovered from the circular vectors were less than 100 bp, and many were less than 10 bp (data not shown). In contrast, cloning the same cDNA fractions into the pJAZZ-OC vector resulted in libraries of clones all within the expected size ranges (Figure 7-4a; Color Addendum Plate 11a). Sequence analysis confirmed that many of these clones consisted of di-, tri-, and tetra-nucleotide repeats (Figure 7-4b; Color Addendum Plate 11b), which presumably led to their deletion in supercoiled vectors.

Efficient Library Construction and Assembly

The pJAZZ vector was used to construct three libraries to sequence the genome of *Flavobacterium columnare* (69% AT, 3.1 Mb). Genomic DNA was randomly sheared, size-selected to 2–6 kb, 6–10 kb, and 10–12 kb, and the fractions were separately cloned into the pJAZZ-KA vector. A total of 88 Mb was cloned, and 33 Mb was sequenced (~10× coverage). Automated assembly proceeded well, following the expected Lander-Waterman curve (Figure 7-5). Initial assembly at 8× genome coverage yielded 21 major contigs of >40 kb, 15 contigs of 2–40 kb, plus 21 small contigs or singletons. Importantly, the gaps between contigs appeared to be short sequencing gaps of <10 kb, rather than physical clone gaps. Therefore, this AT-rich genome appears to have been completely cloned and nearly completely sequenced without the use of BAC or fosmid clones. Gap closure is being carried out by our group.

Cloning of Large PCR Products

PCR amplicons of >10 kb are often generated to span gaps in genomic assemblies or to obtain DNA regions that cannot be cloned by standard methods, for example, phage or viral genomes. To demonstrate the utility of the linear vector in cloning large PCR products, defined sections of the

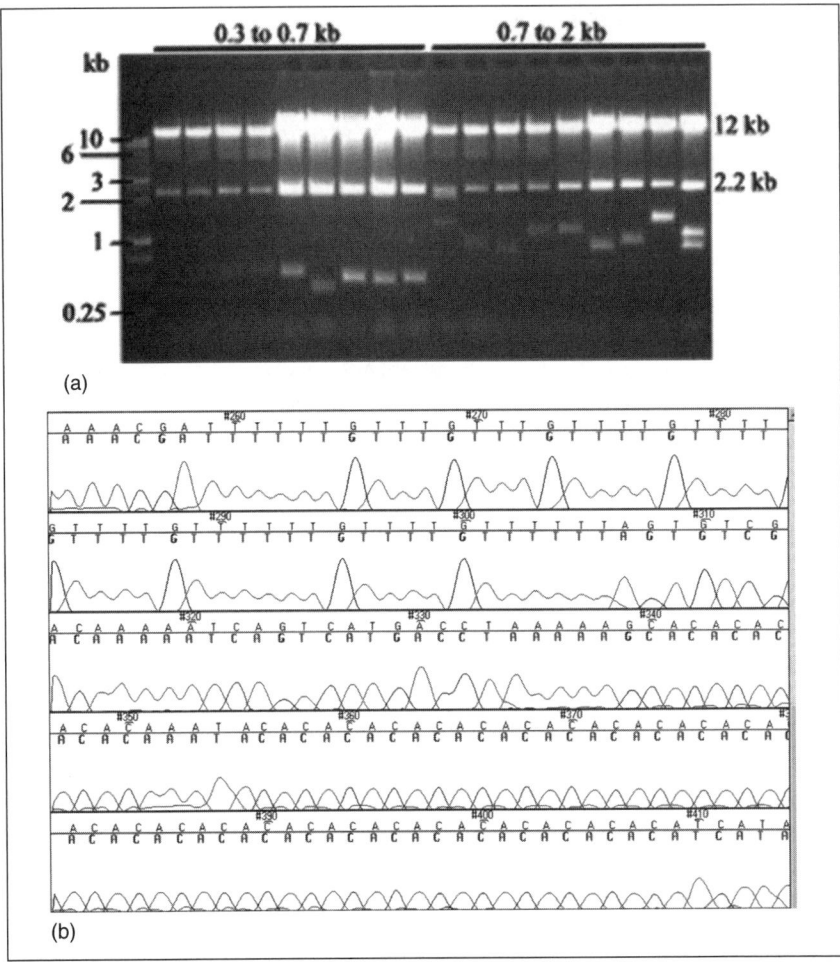

Figure 7-4. Cloning repetitive cDNA in the pJAZZ vector (see Plates 11a and b in the Color Addendum). (a) Clones from libraries of 0.3 to 0.7 kb and 0.7 to 2.0 kb were digested with *Not*I to excise the insert. Nine of the samples were subject to copy number amplification during growth (Lanes 6–10 and 16–19). (b) Chromatogram from sample 18, showing repetitive sequences.

phage lambda genome were amplified and cloned into the pJAZZ vector. Fragments of 15 to 20 kb were routinely cloned in this system, without the need for specialized protocols (Figure 7-6). Clones with inserts of this size were rare with conventional vectors (data not shown).

Lack of Size Bias

With circular vectors, the efficiency of transformation decreases dramatically as the size of plasmid is increased. The effect is evident across all

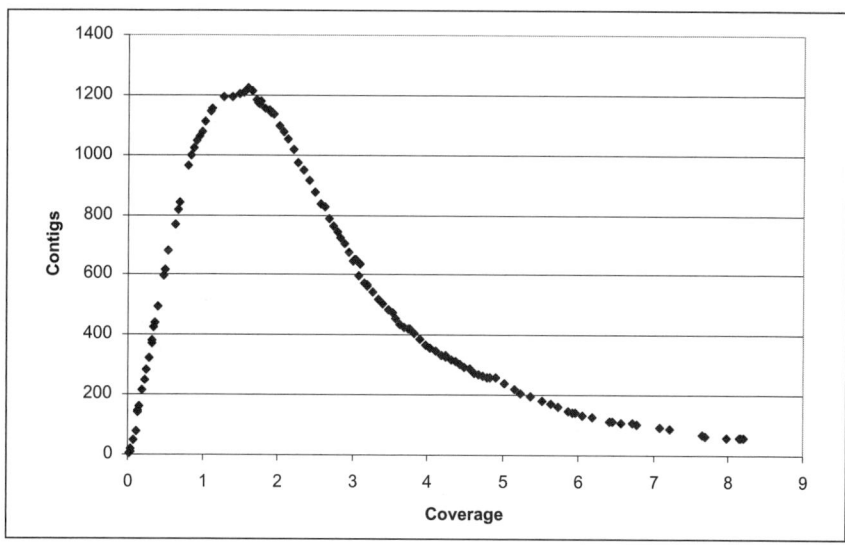

Figure 7-5. Assembly of the *Flavobacterium* genome (69% AT) in the pJAZZ vector. The assembly data closely approximates the predicted Lander Waterman curve.

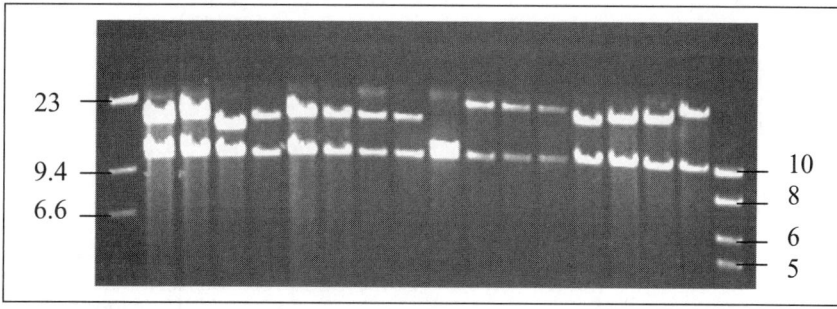

Figure 7-6. Cloning 15 to 20 kb PCR amplicons into the pJAZZ vector. Lambda DNA fragments were amplified with Phusion DNA polymerase (New England Biolabs), phosphorylated with T4 polynucleotide kinase, and cloned into the pJAZZ-OC vector. Clones were digested with *Not*I. Inserts migrate at 10 to 23 kb, and the left vector arm is visible at 10 kb. The right arm has run off the gel.

size ranges of inserts. DNA fragments of 130 kb are cloned five times more effectively than those of 170 kb (47; data not shown). A similar effect is seen with smaller fragments as well. For example, in one trial DNA was sheared to the range of 2 to 6 kb, but it was *not* size-selected further. After end-repair, one aliquot was cloned into the circular vector pSMART-LCKan (2.1 kb), which is a low copy number and transcription-free. An identical aliquot was cloned into the pJAZZ vector. Randomly selected recombinants were predominantly 3 kb in the circular vector, compared to 4 to 5 kb in the pJAZZ vector (Table 7-1).

Table 7-1. Lack of size bias in the pJAZZ vector.

Sheared Fragments			Restriction Fragments		
Insert Size (kb)	pJAZZ	pSMART-LCKan	Insert Size (kb)	pJAZZ	pSMART-VC
6	2	1	17	3	2
5	5	2	12	2	3
4	6	2	8	10	2
3	3	7	7	9	3
2	0	3	3	6	7
<1	2	3	<1	6	19

(Left three columns) Randomly sheared fragments of 2 to 6 kb were ligated into the pJAZZ vector or into the circular vector pSMART-LCKan. (Right three columns) Restriction fragments of 3 to 17 kb were ligated into the pJAZZ or pSMART-VC vectors. Inserts of <1 kb consisted of unexpected fragments of lambda phage DNA (data not shown). Colonies were picked at random for size analysis.

In a second trial, restriction fragments of lambda DNA were cloned into the pJAZZ vector or into the pSMART-VC vector, which is a transcription-free BAC vector of 7 kb. The predominant inserts in the pJAZZ vector were the 7 and 8 kb fragments, whereas the BAC vector mostly contained fragments of 3 kb or <1 kb (Table 7-1).

Size Limit of Inserts

Several sources of very large fragments were ligated into the pJAZZ vector to assess the maximum insert size tolerated. Bacterial or human genomic DNA was digested with *Not*I, size fractionated to 300 kb, and cloned into the pJAZZ vector. Unexpectedly, the maximum insert size appeared to be approximately 40 kb. We also attempted to re-clone BAC inserts of 50, 100, or 150 kb of potato genomic DNA into the pJAZZ vector. As in the previous trial, all clones had inserts in the range of approximtely 30 to 40 kb, regardless of the size of the input DNA (data not shown). These results suggest that there is an inherent size limit of 30 to 40 kb for inserts in this system.

Random Shear BAC Library Construction

Although the linear vector is well suited for large-insert library construction and closing gaps of up to approximately 30 kb, BAC libraries are necessary to obtain larger clones. BAC libraries are typically constructed by partial restriction digestion of genomic DNA, but this method often excludes regions of 10s to 1000s of kb (13, 18, 22). Highly repetitive regions such as telomeres and centromeres may be devoid of restriction sites. Regions with extremely biased GC content may also be under- or over-digested, resulting in their exclusion from the library. In addition, methylation or other modifications may render DNA resistant to cleavage. The resulting clone gaps can be impossible to close with partial digestion BAC libraries, even with multiple complementary restriction digests providing 20× to 40× genome coverage. This phenomenon dramatically increases the effort and cost required for finishing.

Problems due to restriction site bias in the genome can be circumvented by randomly shearing the genomic DNA used for BAC library construction. Initial attempts at random shearing were not successful, presumably due to difficulty in creating and cloning fragments in the correct size range (29). Subsequent success has been reported, generating inserts of 50 to 150 kb, some of which include telomeric regions (29). We have developed alternate methods of random shearing and cloning, generating inserts of >100 kb in the resulting BAC libraries. Using this Random Shear BAC Library, just 5× coverage was sufficient to close several centromeric gaps in the "finished" *Arabidopsis thaliana* genome. We also have developed transcription-free BAC cloning vectors to alleviate additional instability problems.

Random Shearing of Undigestable Genomic DNA

Complete digestion of genomic DNA with common restriction enzymes reveals megabase regions that are resistant to digestion (Figure 7-7, left panel). Hybridization of such regions to centromeric repeats confirms their enrichment for centromeric DNA (data not shown). In contrast, random shearing of the genomic DNA to 100 to 400 kb resulted in complete fragmentation of all material (Figure 7-7, right panel). This method therefore eliminates a major form of bias inherent to preparation of the DNA for BAC cloning.

Cloning Large Inserts in a Random Shear BAC Library

To confirm that randomly sheared genomic DNA was suitable for construction of a large-insert BAC library, potato genomic DNA was randomly sheared, end-repaired, size-selected, and cloned into the pSMART® BAC vector (Lucigen; Figure 7-8a). The average insert size was >100 kb (Figure 7-8b).

To demonstrate that this method results in an unbiased distribution of clones, a Random Shear BAC Library of *Arabidosis* was screened for

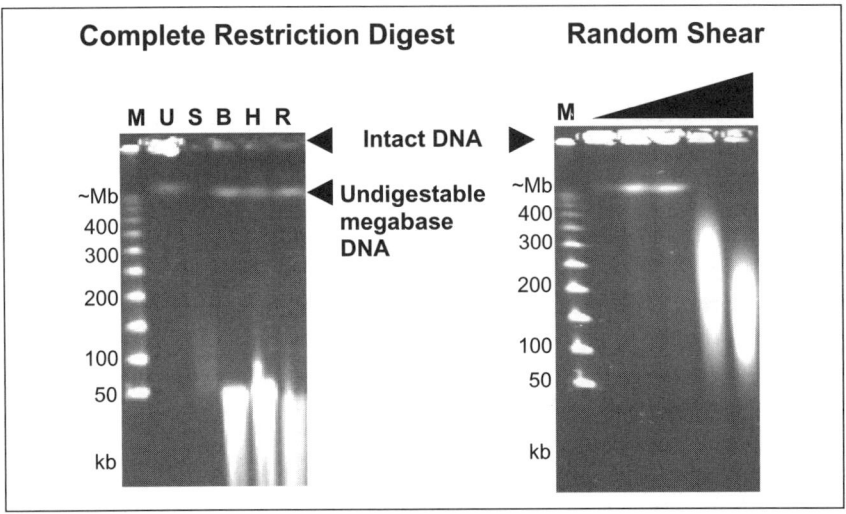

Figure 7-7. **Fragmentation of mouse genomic DNA.** Left panel: Complete digestion by several restriction enzymes reveals undigestable megabase bands. Lanes:
M, Marker; U, Uncut; S, *Sau*3A; B, *Bam*HI; H, *Hind*III; R, *Eco*RI. Intact chromosomal DNA did not enter the gel and is visible in the wells. *Sau*3A digested the
DNA migrating at ~1 Mb, as well as the bulk of the remaining DNA. The majority
of the Sau3A digest ran off the gel, leaving only the faint smear at 50 to 100 kb.
Right panel: Random shearing fragments the Mb-sized DNA along with the bulk
of the remaining DNA.

various regions of the published "finished" genome sequence, derived
from two conventional BAC libraries of *Arabidopsis* (26). These libraries
are equivalent to 17× genomic coverage, but specific regions of Chromosome 1 are strongly under- or over-represented, reflecting the bias of
the partial digestion BAC libraries (46, 48). *Arabidopsis* genomic DNA was
randomly sheared, end-repaired, size-selected, and cloned into the
pSMART BAC vector. A 5× coverage library was screened with oligonucleotide probes for specific regions of Chromosome 1 that were present at
<1, 15, or >75 clones per 0.1 Mb. Significantly, the probes detected 4 to
5 BAC clones from each of these regions in the Random Shear Library
indicating an unbiased distribution of inserts (Figure 7-9). Several of the
Random Shear clones contained previously uncloned centromeric gaps of
this "finished" physical and sequence genomic map. The same probes also
identified clones covering centromeric regions of other chromosomes.
These results suggest that Random Shear BAC Libraries offer a way to
completely finish existing genome sequencing projects.

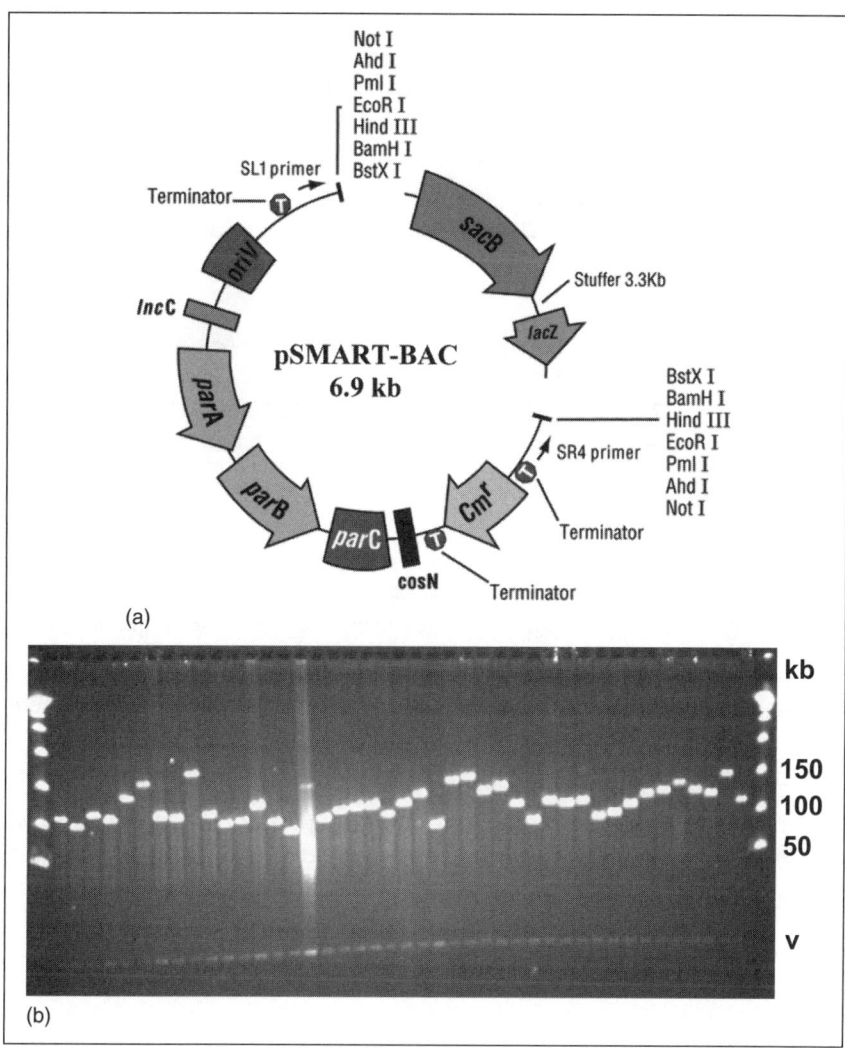

Figure 7-8. **Inserts of more than 100 kb in a Random Shear BAC Library.** Potato genomic DNA was randomly sheared and cloned into the pSMART-BAC vector. (a) Map of the pSMART-BAC vector. (b) Miniprep DNA from recombinants was digested with *Not*I to excise inserts. The vector band is visible at 7 kb ("v").

Amplification of Copy Number in the Random Shear Library

The pSMART BAC vector (Figure 7-9) contains both the single-copy origin of replication from the *E. coli* F-factor and the inducible mid-copy oriV origin of replication. The vector is transformed into the BAC-Optimized Replicator Cells. These cells are an *E. coli* strain that contains the *trfA* gene for initiation of replication from oriV (43). The *trfA* gene is under tight

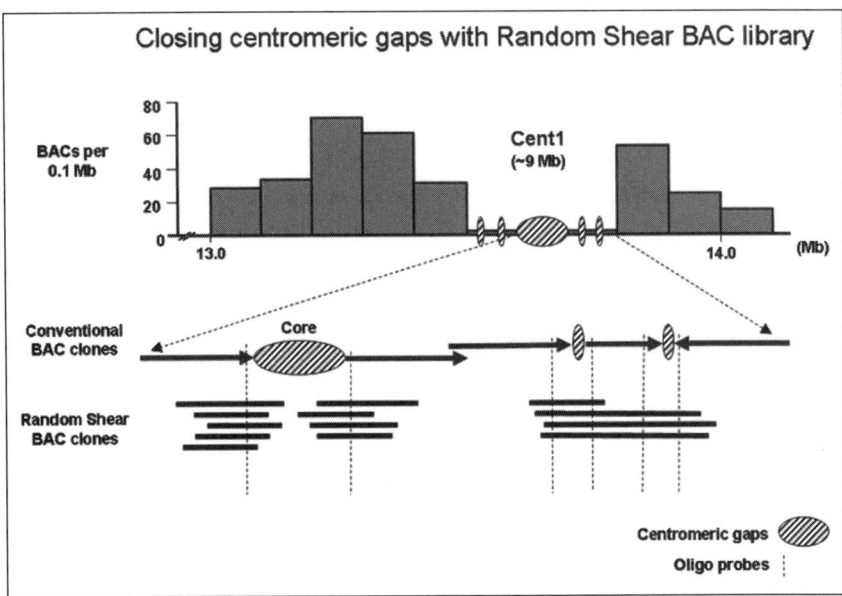

Figure 7-9. Distribution of BAC clones from the *Arabidopsis* genome. This region of chromosome 1 is represented by 0 to >70 clones per Mb in the genome project (26) (bar graph). In contrast, probing with oligonucleotides indicated that the Random Shear Library contained 4 to 5 clones per site, even in centromeric regions.

control of the inducible *ara*BAD promoter. Upon addition of arabinose, pSMART BAC clones replicate to approximately 10 to 20 copies/cell and are stably maintained, providing high yields of high purity DNA (Figure 7-10). This feature simplifies BAC DNA preparation and increases the success rate of BAC end sequencing to >95%.

Summary

We show here that use of pJAZZ linear vectors and Random Shear BAC Libraries can greatly simplify the finishing process in genomic sequencing. The linear vector stably maintains DNA that is "unclonable" in traditional circular vectors, minimizing the number of cloning gaps. The most notable examples were provided by AT-rich inserts of >20 kb or repetitive DNAs of up to 2 kb. Larger fragments of repetitive DNA are currently being investigated in our laboratory to determine the upper size limit for such templates.

The instability of these sequences has not been rigorously explained or understood in all cases, but several factors are known to affect it. In conventional cloning vectors, the high level of transcription from the *lacZ*

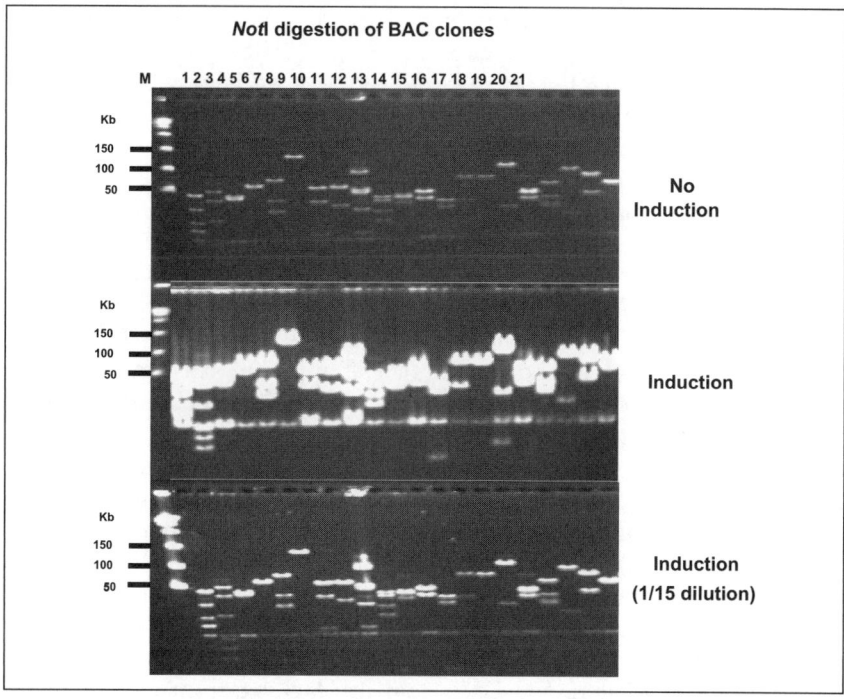

Figure 7-10. **Amplification of pSMART BAC clones.** BAC clones were grown with or without amplification of copy number. BAC DNA was prepared according to standard mini-prep protocols. Upper panel: without induction; middle panel: with induction; lower panel: with induction, diluted 15-fold. Each lane contains 10 µL of DNA, digested with *Not*I. Insert sizes are 50 to ~200 kb. M, lambda DNA ladder.

promoter may generate deletions by any of several mechanisms (12). Further, these vectors typically exist at hundreds of copies per cell, providing ample opportunity for recombination and rearrangement between plasmids. Previous work has demonstrated that these problems can be at least partially alleviated by cloning into a transcription-free, low copy–number vector (12).

However, torsional strain imparted by supercoiling also plays a major role in deletion or rearrangement of unstable sequences. Results obtained with the linear vector corroborate the adverse effects of supercoiling on the stability of repeats (27) and palindromic sequences (21, 24). The AT-rich or repetitive DNAs tested here were deleted or not recovered at all in supercoiled vectors, even in transcription-free or single-copy vectors. But, the same templates were readily cloned into the linear pJAZZ vector. The linear vector also has been shown to tolerate large inverted repeats of 10 to 20 kb (33; data not shown), implying that supercoiling is a major factor in destabilizing such structures.

The pJAZZ system currently is the only method of cloning and maintaining DNAs as linear molecules in *E. coli*. Although bacteriophage lambda vectors are ligated into a linear form for packaging, the molecule is re-circularized for replication. The instability caused by supercoiling is therefore likely to affect inserts in lambda clones.

Using the pJAZZ vector is also simpler, more versatile, and more economical than using lambda-based vectors for cloning large inserts. Ligation, transformation, and preparation of DNA from pJAZZ clones are accomplished with standard methods used for conventional plasmids. The linear vector tolerates any size insert up to approximately 30 to 40 kb, whereas lambda vectors are limited to a narrow size range (30–40 kb for fosmids, 10–15 kb for lambda vectors). The apparent size limit of ~50 kb for the total length of pJAZZ linear clones is not understood. Interestingly, the maximum length tolerated by bacteriophage lambda is also ~50 kb, which is attributed to space limitations inside the phage head. However, the linear vector is not subject to this limitation, suggesting that additional mechanisms limit the size of linear plasmid vectors. Whether such mechanisms also play a role in native linear phages is unknown.

In conjunction with the linear vector, the use of Random Shear BAC Libraries can contribute greatly to the finishing efforts in genome sequencing projects. A single Random Shear BAC Library appears to include regions of genomic DNA that are lacking from assemblies of multiple conventional partial-digest libraries. The reduced cloning bias can provide access to previously unclonable regions, with a minimal number of redundant clones. The results suggest that randomly sheared BAC libraries can yield superior data, while greatly decreasing the required time, effort, and expense (Table 7-2).

Table 7-2. Comparison of Random Shear and partial digestion BAC Libraries.

BAC Library	Character	Number of Libraries Needed (coverage)	Cost of Library Construction* ($US)	Finishing Cost** ($US)
Random Shear	Unbiased; no gaps	1 (10×)	<$30,000	<$1 million
Partial Digestion	Biased; gaps	≥2 (>20×)	≥$60,000	≥$2 million

* Construction cost is based on a genome of ~500 Mb, using either one Random Shear BAC Library (10× coverage) or at least two complementary partial-digest BAC Libraries to minimize restriction site bias (20× total coverage).
** Finishing cost includes BAC end sequencing, whole genome physical mapping, and integrating the physical map with ~1000 genetic makers.

References

1. Adhya, S., and Gottesman, M. 1982. Promoter occlusion: transcription through a promoter may inhibit its activity. *Cell* 29: 939–944.
2. Bichara, M., Pinet, I., Schumacher, S., et al. 2000. Mechanisms of dinucleotide repeat instability in *Escherichia coli*. *Genetics* 154: 533–542.
3. Bowater, R.P., Jaworski, A., Larson, J.E., et al. 1997. Transcription increases the deletion frequency of long CTG·CAG triplet repeats from plasmids in *Escherichia coli*. *Nucleic Acids Res* 25: 2861–2868.
4. Celniker, S.E., Wheeler, D.A., Kronmiller, B., et al. 2002. Finishing a whole-genome shotgun: release 3 of the Drosophila melanogaster euchromatic genome sequence. *Genome Biology* 3(12): research0079.1–0079.14. Available at http://genomebiology.com/2002/3/12/research/0079.
5. Cunningham, T.P., Montelaro, R.C., and Rushlow, K.E. 1993. Lentivirus envelope sequences and proviral genomes are stabilized in *Escherichia coli* when cloned in low-copy-number plasmid vectors. *Gene* 124: 93–98.
6. Deneke, J., Ziegenlin, G., Lurz, R., et al. 2000. The protelomerase of temperate *Escherichia coli* phage N15 has cleaving-joining activity. *Proc Natl Acad Sci U S A* 97: 7721–7726.
7. Doak, T., Cavalcanti, A.R., Stover, N.A., et al. 2003. Sequencing the Oxytricha trifallax macronuclear genome: a pilot project. *Trends Genet* 19: 603–607.
8. Dorokhov, B.D., Lane, D., and Ravin, N.V. 2003. Partition operon expression in the linear plasmid prophage N15 is controlled by both Sop proteins and protelomerase. *Mol Microbiol* 50: 713–721.
9. Feng, T., Li, Z., Jiang, W., et al. 2002. Increased efficiency of cloning large DNA fragments using a lower copy number plasmid. *Biotechniques* 32: 992–996.
10. Gardner, M.J., Hall, N., Fung, E., et al. 2002. Genome sequence of the human malaria parasite *Plasmodium falciparum*. *Nature* 419: 498–511 (see Comments).
11. Glockner, G., Eichinger L., Szafranski, K., et al. 2002. Sequence and analysis of chromosome 2 of *Dictyostelium discoideum*. *Nature* 418: 79–85.
12. Godiska, R., Patterson, M., Schoenfeld, T., et al. 2005. *Beyond pUC: Vectors for Cloning Unstable DNA*. In: *DNA Sequencing: Optimizing the Process and Analysis*. Kieleczawa, J., ed. Sudbury, MA: Jones and Bartlett.
13. Hosouchi, T., Kumekawa, N., Tsuruoka, H., et al. 2002. Physical map-based sizes of the centromeric regions of *Arabidopsis thaliana* chromosomes 1, 2, and 3. *DNA Res* 9: 117–121.
14. National Human Genome Research Institute, National Institutes of Health. *International Consortium Completes Human Genome Project*. Available at http://www.genome.gov/11006929; accessed November 5, 2007.
15. International Human Genome Sequencing Consortium. 2004. Finishing the euchromatic sequence of the human genome. *Nature* 431: 931–945.
16. Jaworski, A., Rosche, W.A., Gellibolian, R., et al. 1995. Mismatch repair in *Escherichia coli* enhances instability in vivo of $(CTG)_n$ triplet repeats from human hereditary diseases. *Proc Natl Acad Sci U S A* 92: 11019–11023.

17. Kiyama, R., and Oishi, M. 1994. Instability of plasmid DNA maintenance caused by transcription of poly(dT)-containing sequences in *Escherichia coli. Gene* 150: 57–61.
18. Kouprina, N., Leem, S.H., Solomon, G., et al. 2003. Segments missing from the draft human genome sequence can be isolated by TAR cloning in yeast. *EMBO Rep* 4: 257–262.
19. Kunst, F., Ogasawara, N., Moszer, I., et al. 1997. The complete genome sequence of the gram-positive bacterium *Bacillus subtilis. Nature* 390: 249–256.
20. Lander, E.S., Linton, L.M., Birren, B., et al. 2001. Initial sequencing and analysis of the human genome. *Nature* 409: 860–921 (see Comments).
21. Leach, D., and Lindsey, J. 1986. *In vivo* loss of supercoiled DNA carrying a palindromic sequence. *Mol Gen Genet* 204: 322–327.
22. Leem, S.H., Kouprina, N., Grimwood, J., et al. 2004. Closing the gaps on human chromosome 19 revealed genes with a high density of repetitive tandemly arrayed elements. *Genome Res* 14:239–246.
23. Lyer, R.R., and Wells, R.D. 1999. Expansion and deletion of triplet repeat sequences in *Escherichia coli* occur on the leading strand of DNA replication. *J Biol Chem* 274: 3865–3877.
24. Malagon, F., and Aguilera, A. 1998. Genetic stability and DNA rearrangements associated with a 2 × 1.1-Kb perfect palindrome in *Escherichia coli. Mol Gen Genet* 259: 639–644.
25. Mardanov, A.V., Strakhova, T.S., and Ravin, N.V. 2006. Functional characterization of the repA replication gene of linear plasmid prophage N15. *Res Microbiol* 157: 176–183.
26. Mozo, T., Dewar, K., Dunn, P., et al. 1999. A complete BAC-based physical map of the Arabidopsis thaliana genome. *Nat Genet* 22: 271–275.
27. Napierala, M., Bacolla, A., and Wells, R.D. 2005. Increased negative superhelical density *in vivo* enhances the genetic instability of triplet repeat sequences. *J Biol Chem* 280: 37366–37376.
28. Ohshima, K., Kang, S., Larson, J.E., et al. 1996. Cloning, characterization, and properties of seven triplet repeat DNA sequences. *J Biol Chem* 271: 16773–16783.
29. Osoegawa, K., Vessere, G.M., Shu, C.L., et al. 2007. BAC clones generated from sheared DNA. *Genomics* 89: 291–299.
30. Parniewski, P., Bacolla, A., Jaworski, A., et al. 1999. Nucleotide excision repair affects the stability of long transcribed (CTG·CAG) tracts in an orientation-dependent manner in *Escherichia coli. Nucleic Acids Res* 27: 616–623.
31. Ravin, N., and Lane, D. 1999. Partition of the linear plasmid N15: interactions of N15 partition functions with the sop locus of the F plasmid. *J Bacteriol* 181: 6898–6906.
32. Ravin, N.V., Kuprianov, V.V., Gilcrease, E.B., et al. 2003. Bidirectional replication from an internal ori site of the linear N15 plasmid prophage. *Nucleic Acids Res* 31: 6552–6560.
33. Ravin, N.V., and Ravin, V.K. 1998. Cloning of large imperfect palindromes in circular and linear vectors. *Russian Journal of Genetics* 34: 38–44.
34. Ravin, N.V., and Ravin, V.K. 1999. Use of a liner multicopy vector based on the mini-replicon of temperate coliphage N15 for cloning DNA with abnormal secondary structures. *Nucleic Acids Res* 27: e13i–e13iii.

35. Ravin, N.V., Strakhova, T.S., and Kuprianov, V.V. 2001. The protelomerase of the phage-plasmid N15 is responsible for its maintenance in linear form. *J Mol Biol* 312: 899–906.

36. Ravin, N.V. 2003. Mechanisms of replication and telomere resolution of the linear plasmid prophage N15. *FEMS Microbiol Lett* 221: 1–6

37. Ravin, V., Ravin, N., Casjens, S., et al. 2000. Genomic sequence and analysis of the atypical temperate bacteriophage N15. *J Mol Biol* 299: 53–73.

38. Rybchin, V.N., and Svarchevsky, A.N. 1999. The plasmid prophage N15: a linear DNA with covalently closed ends. *Mol Microbiol* 33: 895–903.

39. She, X., Jiang, Z., Clark, R.A., et al. 2004. Shotgun sequence assembly and recent segmental duplications within the human genome. *Nature* 431: 927–930.

40. Stueber, D., and Bujard, H. 1982. Transcription from efficient promoters can interfere with plasmid replication and diminish expression of plasmid specified genes. *EMBO J* 1: 1399–1404.

41. Tilly, K. 1991. Independence of bacteriophage N15 lytic and linear plasmid replication from the heat shock proteins DnaJ, DnaK, and GrpE. *J Bact* 173: 6639–6642.

42. Venter, J.C., Adams, M.D., Myers, E.W., et al. 2001. The sequence of the human genome. *Science* 291: 1304–1351.

43. Wild, J., Hradecna, Z., and Szybalski, W. 2002. Conditionally amplifiable BACs: switching from single-copy to high-copy vectors and genomic clones. *Genome Res* 12: 1434–1444.

44. Wu, C., Santos, F., Nimmakayala, P., et al. 2004. Construction of a complementary BAC library for soybean genome-wide physical mapping and whole genome sequencing. *Theor Appl Genet* 109: 1041–1050.

45. Wu, C., Sun, S., Lee, M.K., et al. Whole genome physical mapping: an overview on methods for DNA fingerprinting. In: Meksem K. and Kahl G., eds. *The Handbook of Plant Genome Mapping: Genetic and Physical Mapping.* Weinheim, Germany: Wiley-VCH Verlag GmbH; 2005: 257–284.

46. Wu, C., Xu, Z., and Zhang, H.B. DNA Libraries. In: Meyers R.A., ed. *Encyclopedia of Molecular Cell Biology and Molecular Medicine,* vol 3, 2nd ed. Weinheim, Germany: Wiley-VCH Verlag GmbH; 2004: 385–425.

47. TAIR. The Arabidopsis Information Resource. Columbus, OH: Ohio State University. Available at http://www.arabidopsis.org; accessed November 5, 2007.

48. Zhang, H.B., and Wu, C. 2001. BAC as tools for genome sequencing. *Plant Physiol Biochem* 39: 195–209.

8 — Bioinformatics Tools to Aid Sequencing of Difficult Templates

Jan Kieleczawa, Bharath Lakshmanan,
Donald Koffman, and Aaron Kitzmiller
Wyeth Research, Cambridge, MA

The Sanger DNA sequencing method (19) is a well-established and mature technology that revolutionized our understanding of biology and medicine (10, 21). Over the past few years, more than five hundred genomes of microorganisms and various higher species were completely sequenced and about 2000 additional genomes are in various stages of completion (for the current completed list and status of ongoing sequencing efforts please visit, www.tigr.org). The predominant methodology to sequence all of those organisms was (and still is) based on a shotgun approach where the genomic DNA is split into smaller fragments (e.g., 2, 10 kbp and sometimes BACs) followed by end sequencing. The vector portions of individual reads are removed and "clean" sequences assembled into large contiguous fragments. Any remaining gaps in coverage are most of the time hand curated using a variety of approaches (PCR, alternative chemistry, additional library, etc.). As evidenced by the number of completed organisms, such an approach is extremely effective. The addition of new sequencing technologies (454-Roche, Illumina, AB, Helicos, IBS, and others soon to be on the market) will only speed up this process and make it even more cost effective. Currently, the sequencing of a whole organism requires a huge infrastructure consisting of (at least) library preparation efforts, a significant number of automated sequencing instruments, and bioinformatics tools, and is conducted in a few large genome-sequencing centers. This certainly is changing as even one new generation sequencing instrument is capable of sequencing/re-sequencing a whole genome in a relatively short period of time at a very affordable cost.

On the other hand there are thousands of core DNA sequencing facilities (either stand alone or as a part of a bigger core facility) where the primary effort is to fully sequence/confirm the sequence of relatively few genes, generate single-pass end reads, or any combination thereof. It is worth noting that a DNA sequencing core facility may operate in many different manners. We can classify manner of operation of core facilities into the following levels:

A. *Level 1.* The core receives sequenced and purified samples and the role of the facility is to run samples on sequencing instruments and deliver raw, unedited chromatograms. The submitter is responsible for making sure that the amount of DNA, primer, cycling conditions, and cleanup of sequencing reactions are carried out under optimal conditions. Analysis of the data and the decision to add more reactions using modified sequencing protocols (e.g., when encountering a difficult region) is in the hands of a requestor. The core facility staff does not need to have any specialized knowledge related to the DNA sequencing and needs only to be well trained on usage and maintenance of sequencing instruments. This training can be easily accomplished within just a few days.

B. *Level 2.* The core receives template DNAs mixed with primers (1DNA/1 primer per tube or well, depending on the sample submission requirements). The core performs cycle sequencing, cleanup, and delivers typically raw, unedited data. Some level of analysis could be part of results delivery as agreed between a core and a specific user. A decision to add new reactions or modified conditions most often will rest with the submitter, although in some cases designated staff at core facility can assume full responsibility for delivering finished data, for instance, fully double-stranded coverage as specified by a requestor. This process will involve designing and ordering new primers (if needed), data editing, and possibly comparison to a designated reference sequence. Some type of a basic report (no or incorrect insert, no mutation, one mutation, many mutations compared to reference sequence or no assembly to reference sequence, etc.) could be part of the operation. This could help requestors to concentrate on templates with the desired characteristics (core facility staff may or may not be aware what a requestor is looking for or even if the provided reference sequence is correct or not).

The core facility staff not only needs to be trained as in the level facility 1, but all or some personnel must be familiar with all other steps needed for efficient running of a sequencing facility. In addition, familiarity with editing, assembly, and interpretation of data is required for the portion of operation where staff is asked to deliver finished data.

C. *Level 3.* The core receives either colonies, glycerol stock, or prepared DNA and some basic information/instruction like the vector the clone is in, desired range to sequence, insert size, the depth of coverage (end reads, fully single or double-stranded coverage, gap closing, etc.), reference sequence or NCBI's accession number, and any other data that may be useful to complete a project. The staff of the core facility is fully responsible for assigning the correct vector primers and/or selecting or designing internal primers to accomplish the stated task. If needed, they decide what modified conditions to use to get through any difficult region. They also perform data editing and analysis with comments pertaining to each analyzed clone (see description in Level 2).

Almost all members of the core facility (with the exception of person(s) who handle cleanup, maintenance of instruments, and any other routine tasks that are specific to running of the facility) need very extensive knowledge in dealing with all sorts of templates and processes. These include: the amount of DNA and primer needed for sequencing; volume of dye-terminator; cycling protocols; modified sequencing protocols to read-through many kinds of difficult templates; parameters for primer design; and data editing/assembly using specialized programs like Sequencher, Consed, DNAStar, and/or others. However, if the core facility staff is sophisticated, some sequencing parameters could be incorporated into a customized LIMS and all calculations could be performed based on pre-determined default values (4D LIMS sequencing database has many such capabilities as described below and in reference 18).

To be highly efficient and successful, the core facility at Level 3 (and to some degree at Level 2) needs to have in place at least four elements:

1. Knowledgeable and experienced staff.
2. Robust and flexible sequencing and auxiliary protocols (like setup of sequencing reactions and their cleanup prior to loading onto a sequencing instrument, cycling protocols, primer design, etc.).
3. Reliable instrumentation (DNA sequencers, thermocyclers, robotics).
4. Customized LIMS, editing/assembly program and bioinformatics tools to analyze provided reference sequences to detect potential difficult regions, to quickly design primers and to compare with "real sequences."

Over the years, we have built all four components into our DNA sequencing core facility at Wyeth Research. Our staff averages over 13 years' experience in sequence finishing and we have developed new or improved all existing sequencing protocols (11–17). Currently, we are using state-of-the-art capillary sequencing instrumentation (ABI 3730;

Applied Biosystems, Foster City, CA), robotics (Biomek NX; Beckman Coulter, Fullerton, CA; and Hamilton Starlet; Hamilton, Reno, NV), and other auxiliary instrumentation. We are using the Sequencher program (Gene Codes, Ann Arbor, MI) for data editing and analysis, an ideal, in our view, software tool for our operation where all sequences need to be base-perfect, fully double stranded. In the last few years, we have developed and now continually add new functionality to our custom DNA sequencing LIMS. The general features of this LIMS were described earlier (18).

In this chapter, we describe in detail two new features of the DNA Sequencing LIMS specifically designed to improve the processing of various difficult templates. The first component (GC module) is an integral part of the system and is designed to calculate GC percentage for the provided reference sequence and assign specific chemistry based on the GC content. The second component (Examine Repeats module) calculates up to seven different sequence motifs that typically cause problems while sequencing using standard sequencing protocols (1, 17).

GC Module

Based on the data in Table 2-1 and other referenced publications (2–4, 6), GC-rich regions seem to be the most commonly encountered difficult templates. The most common way to deal with GC-rich regions is to add DMSO to the sequencing reaction mix. In our experience, the addition of DMSO (2.5% to 5% final concentration) sometimes is effective for regions with a GC content less than 70%, but adding just a heat denaturation step is even more effective (8; see Chapter 2). When the GC content of the DNA template (or just a portion of the template) exceeds 70%, the addition of DMSO is rarely effective and other approaches are needed to get through these difficult stretches (11–13, 17). Based on previously published work (11, 17) and the data presented in this chapter, we can suggest the following approach while sequencing templates with different levels of GC content:

1. *GC content ≤70%*. Just add 3 to 5 minutes of the heat denaturation step to a standard ABI-like DNA sequencing protocol (1). The optimal heat denaturation time depends on the lengths of the plasmid DNA (13). PCR fragments can be heat denatured for up to 10 minutes and the sequence quality is not affected much (see Chapter 1).
2. *GC content of a template (or portion of it) is between 71% and 80%*. Use betaine (from Sigma-Aldrich, St. Louis, MO) or reagent A (from Invitrogen, Inc., Carlsbad, CA).

3. *GC content is between 81% and 90%.* Use dGTP V 3.0 sequencing chemistry (substitute BigDye™ BD3.1 with dGTP 3.0 chemistry; one needs to be aware of compression issues). It may be necessary to combine dGTP and regular chemistry to make the correct base calls.

4. *GC content is ≤90%.* Either deaza-7-dGTP PCR/sequencing (11) or Templiphi approach (20; see Chapter 4) may need to be implemented to successfully sequence through such a difficult region.

The suggested chemistries for the above-proposed GC ranges will work most of the time. However, the sequences that precede and follow the difficult stretch are most likely as important in successful read-through of such a difficult stretch (J.K., unpublished observation). Clearly, many different examples of successful sequencing through various difficult stretches and the detailed sequence analysis are needed to develop more generalized rules and to increase the probability of generating good quality data without a try-and-repeat-and-repeat approach.

To deal more effectively with templates containing GC-rich regions, we have integrated a GC calculation module into our LIMS (17, 18). This module allows detailed display of GC content of the DNA template/reference sequence in any range defined by the user (the default interval is about 500 bases) and to automatically assign appropriate sequencing chemistry for the selected primer in this interval. Figure 8-1 shows the GC-content of a DNA molecule with default ranges: none of the regions exceeds 70% GC and no additive is suggested by the LIMS to sequence through this template (Figure 8-2). However, when the GC interval is changed to 200 bases (now GC content exceeds 70% in two intervals; Figure 8-3) and if the needed primer is in one of these intervals, the LIMS software modifies the sequencing reaction by suggesting the addition of betaine to successfully complete the sequencing project (Figure 8-4). The sequencer has the option to override this suggestion, but otherwise the system will recalculate volumes of all reaction components based on the default final volume and other default reaction volumes (DNA concentration, primer, Taq mix and $TE_{sl} = 10\,mM\,Tris/0.01\,mM\,EDTA, pH\,8$). Figure 8-5 shows the DNA molecule with GC content over 80% and consequently another type of modified chemistry will be suggested to sequence through such a region (Figure 8-6).

Examine Repeats Module

Besides GC-rich regions, other forms of repeats and low complexity sequence cause problems with sequencing reactions. Common genomic forms of low complexity (e.g., CA, alu repeats) can cause problems, but

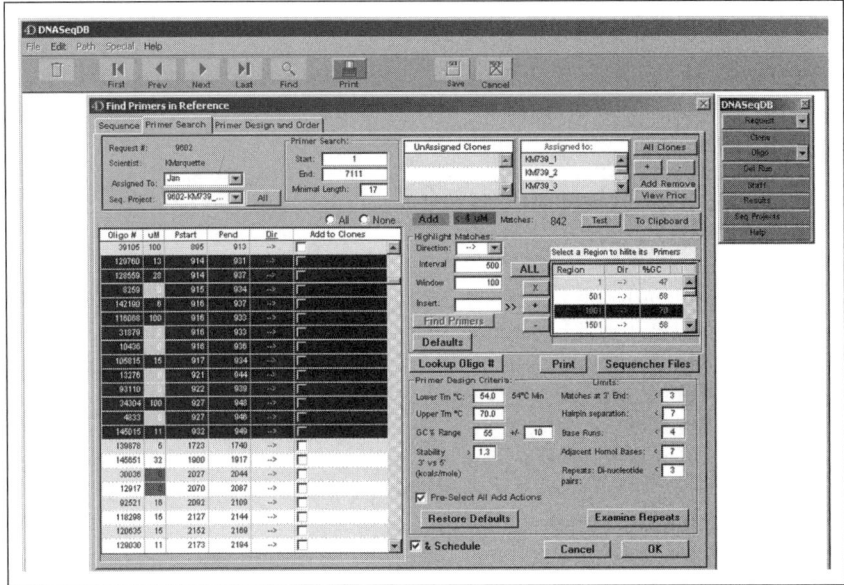

Figure 8-1. The GC content in the reference sequence of 7111 bases is displayed in intervals of 500 bases (default value). None of the shown intervals has a GC content over 70% and, therefore, no special chemistry is suggested. By highlighting the "1000" line in the "Select a Region to hilite its Primers" box (middle right), all primers that prime (forward direction) in the range of 500 to 1000 bases are marked and displayed (left). The primers that prime in reverse direction are at the bottom and are not visible due to the number of over 800 primers found that can be used in our current library. The range of the region for displaying the GC content is customizable by typing any number in the "Interval" box (middle). The default parameters for designing new primers are shown in the right lower part of this panel. These default parameters are also customizable but the user is prevented from selecting "unreasonable" values (e.g., four Cs or Gc in the row).

genomic projects are usually easy for Level 3 core facilities to identify and prepare for ahead of the sequencing process. In addition, mature tools such as RepeatMasker have been available for some time to help identify these types of repeats in template sequences. More problematic has been the increasing use of RNAi technologies (5, 7, 9) containing inverted repeats and the sequencing failures associated with such secondary/tertiary structures (12, 17). Also troublesome are odd forms of low complexity, such as CA dinucleotide repeats, trinucleotide repeats, homopolymers, and dinucleotide nonrepeats (a long stretch of two bases that are not in a consistent order). Because of the emphasis of repeat masking tools on genomic data and database searches, these specific problems for sequencing facilities require new tools. These types of repeats account for as much as 3% to 5% of sequencing templates in our laboratory.

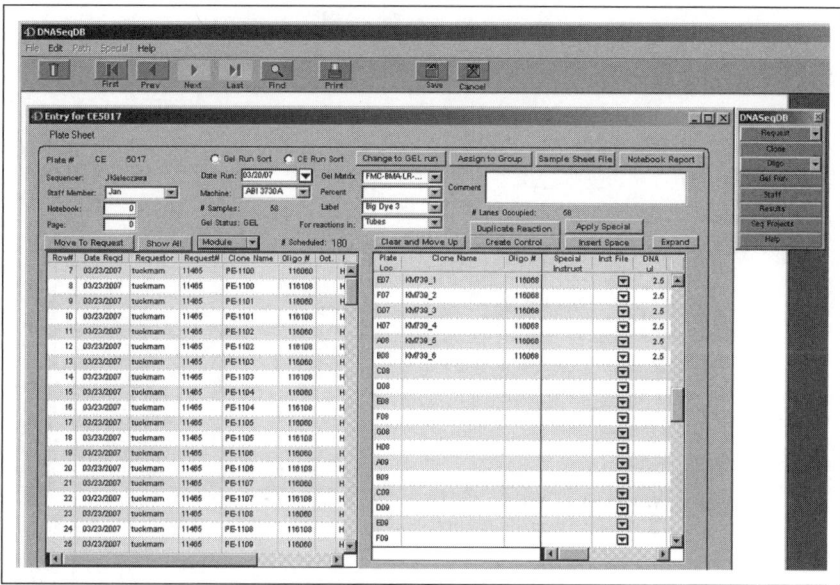

Figure 8-2. The primer 116068 (refers to the sequential primer number in our library) was selected for sequencing on selected clones and scheduled for capillary run. Notice that the "Special instruction" cell on the right hand side of this figure is empty as no special chemistry is suggested to sequence through this level of GC content.

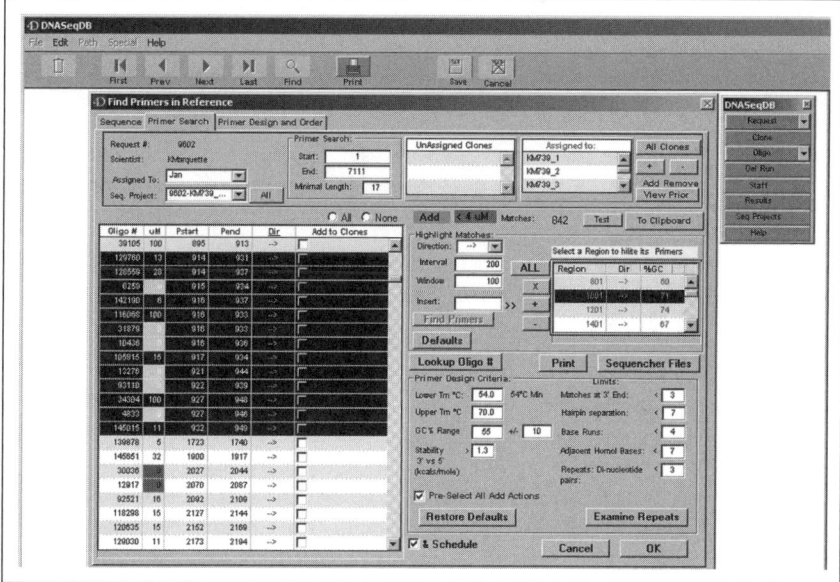

Figure 8-3. The GC content in the reference sequence of 7111 bases (same as in Figure 8-1) is displayed in intervals of 200 bases. This time the GC content in one of the intervals exceeds 70% and if the sequencing primer is selected in this region, the 4D LIMS will suggest a modified chemistry (as shown in Figure 8-4).

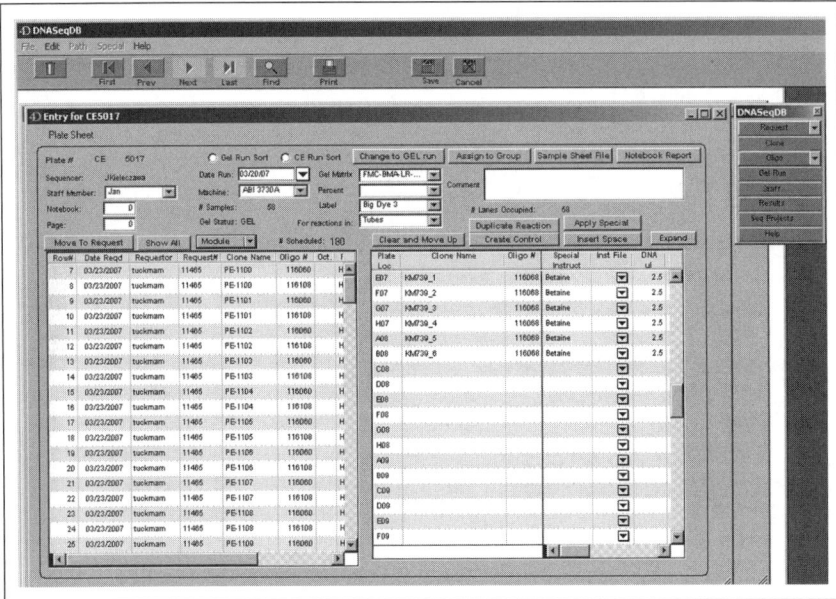

Figure 8-4. The modified sequencing chemistry is suggested when the GC content exceeds 70%. As the GC content is lower than 80% the addition of betaine is suggested to help with sequencing through this region. The user has the option to override this selection by clicking on "Apply Special" button with the list of other choices (not shown). The database automatically recalculates volumes of TE$_{sl}$ buffer and the additive based on default values. The volumes of DNA and dye-terminator are not modified.

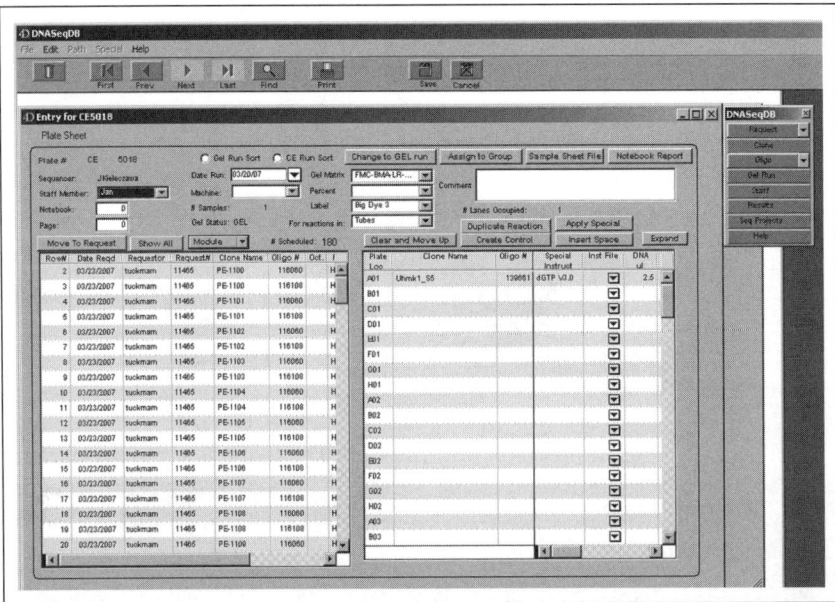

Figure 8-5. The GC content in the reference sequence of 5812 bases displayed in intervals of 200 bases. The GC content in one of the intervals exceeds 80% and therefore different types of special sequencing chemistry is suggested to get clean read through as seen in Figure 8-6.

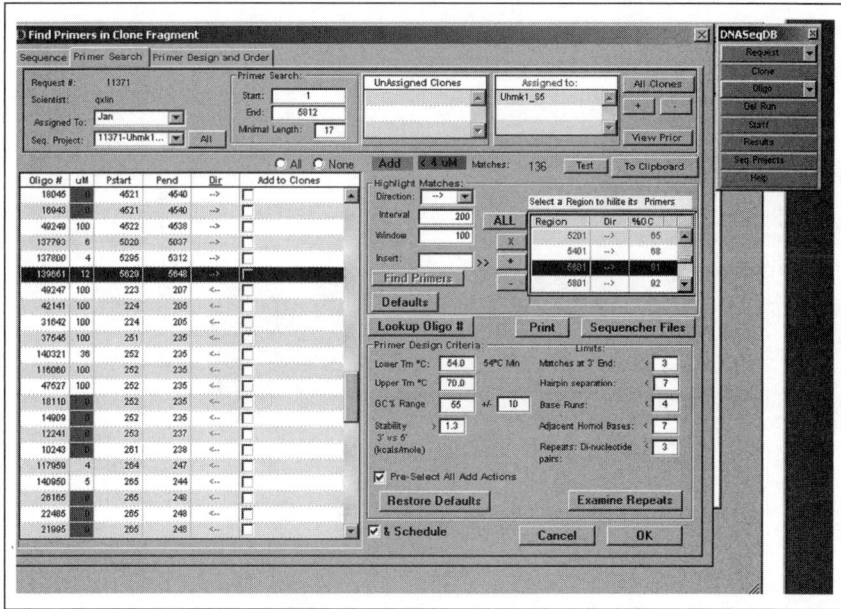

Figure 8-6. **The different modified sequencing chemistry is suggested when the GC content exceeds 80%.** When the GC content is over 80% (but less than 90%) using dGTP instead of regular big dye-terminator is suggested to help with sequencing through this region. The user should be, however, aware of the limitations while using this chemistry as the compressions could obscure true sequence in some regions.

The Examine Repeats module developed by Wyeth Bioinformatics consists of two components covering seven repeat types: direct and inverted repeats, palindromes, di- and tri-nucleotide repeats, dinucleotide non-repeats, homopolymers, and motifs causing compressions. The LIMS request component (Figure 8-7) allows the length and threshold to be custom set for each category, but default values are suggested based on experimental data collected in our laboratory while sequencing many different types of difficult templates (see Chapter 2). The triage repeat finder performs the regular expression searches required to detect repeats in the reference sequence.

The LIMS user interface component allows sequencers to set parameters and displays results (examples shown in Figures 8-8 to 8-10). Integration with the Linux-based Triage Perl CGI code is achieved via a REST-style (8) remote execution.

The Triage repeat finder handles direct and inverted repeats in a similar fashion. For a given template, a segment of sequence—varying between a set minimum (4 bases) and half the total length—is extracted

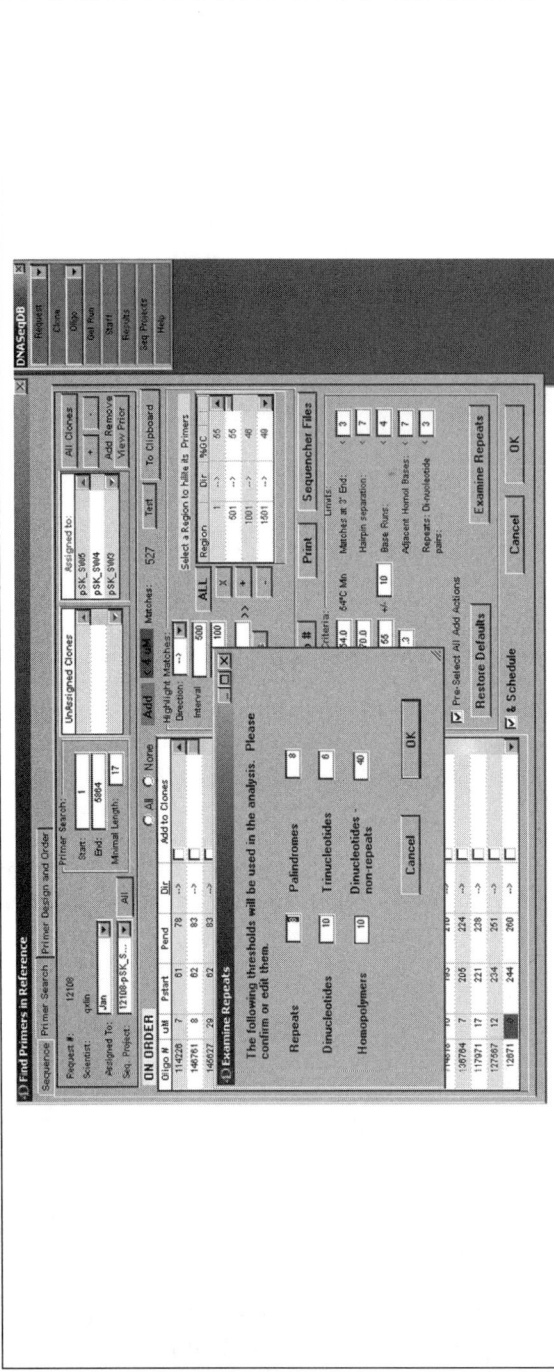

Figure 8-7. The default parameters for Examine Repeats module. A user has an option to use the default values or decrease/increase them as needed. In most cases, modified chemistries will be needed if the length of a repeat or other difficult region exceeds the default value. Note that choice referring to "motifs causing compressions" is not visible in the default window; however, this motif will be displayed if encountered in the sequence (not shown).

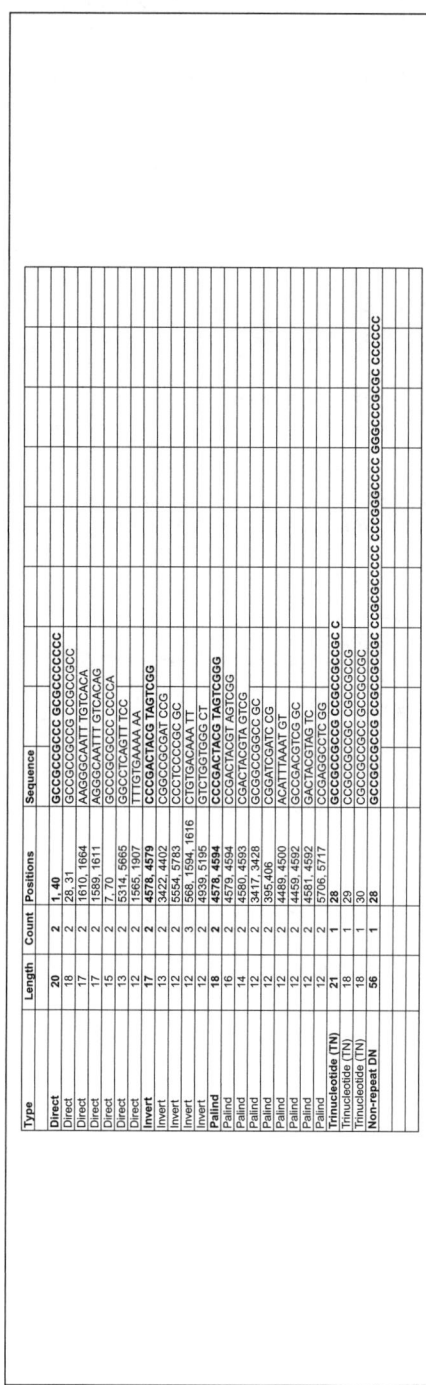

Type	Length	Count	Positions	Sequence
Direct	**20**	**2**	**1, 40**	**GCCGCCGCCC GCGCCCCCC**
Direct	18	2	28, 31	GCCGCCGCCG CCGGCCGCC
Direct	17	2	1610, 1664	AAGGGCAATT TGTCACA
Direct	17	2	1589, 1611	AAGGGCAATT GTCACAG
Direct	15	2	7, 70	GCCGCCGCCG CCCCA
Direct	13	2	5314, 5665	GGCCTCAGTT TCC
Direct	12	2	1565, 1907	TTTGTGAAA AA
Invert	**17**	**2**	**4578, 4579**	**CCCGACTACG TAGTCGG**
Invert	13	2	3422, 4402	CGGGCGCGAT CCG
Invert	12	2	5554, 5783	CCCTCCCCGC GC
Invert	12	3	568, 1594, 1616	CTGTGACAAA TT
Invert	12	2	4939, 5195	GTCTGGTGGG CT
Palind	**18**	**2**	**4578, 4594**	**CCCGACTACG TAGTCGGG**
Palind	16	2	4579, 4594	CCGACTACGT AGTCGG
Palind	14	2	4580, 4593	CGACTACGTA GTCG
Palind	12	2	3417, 3428	GCGGCCGGCC GC
Palind	12	2	395, 406	CGGATCGATC CG
Palind	12	2	4489, 4500	ACATTAAAT GT
Palind	12	2	4459, 4592	GCCGACGTCG GC
Palind	12	2	4581, 4592	GACTACGTAG TC
Palind	12	2	5706, 5717	CCGAGGCCTC GG
Trinucleotide (TN)	**21**	**1**	**28**	**GCCGCCGCCG CCGCCCCGGC C**
Trinucleotide (TN)	18	1	29	CCGCCGCCGC CGCCGGCG
Trinucleotide (TN)	18	1	30	CCGCCGCCGC CGCCGCGCG
Non-repeat DN	**56**	**1**	**28**	**GCCGCCGCCG CCGCCGCCGC CGCGGCCCCC CCCGGGGCCCCC CCGCGCCCCC GGGCCCGCGC CCCCC**

Figure 8-8. **Example of the output from the Examine Repeats module.** The longest difficult motifs in each category are bolded. This DNA molecule is very difficult to sequence as it contains very high GC content (over 95% over stretch of about 150 bases) and in addition it contains tri-nucleotides and long GC non-repeat stretch. Other potential difficult motifs should not interfere with obtaining clean sequencing traces.

Type	Length	Count	Positions	Sequence	
Direct	13	2	103,111	GATCTATAGA TCT	
Direct	13	2	827,841	GAATAGGAAC TTC	
Direct	13	2	327,344	CGGCCATTTA AAT	
Invert	**20**	**2**	**741,768**	**GAGATGAATC AGAGCAGACT**	
Invert	19	2	103,104	GATCTATAGA TCTATAGAT	
Invert	13	2	110,111	AGATCTATAG ATC	
Invert	13	2	804,829	CCGAAGTTCC TAT	
Invert	11	3	806,829,843	GAAGTTCCTA T	
Palind	20	2	103,122	GATCTATAGA TCTATAGATC	
Palind	18	2	104,121	ATCTATAGAT CTATAGAT	
Palind	16	2	105,120	TCTATAGATC TATAGA	
Palind	14	2	110,123	AGATCTATAG ATCT	
Palind	14	2	106,119	CTATAGATCT ATAG	
Palind	12	4	103,114,111,122	GATCTATAGA TC	
Palind	12	2	107,118	TATAGATCTA TA	

Figure 8-9. The Examine Repeat module detected the presence of 20-base inverted repeat structure (bolded) with seven-base loop. The need to use standard or some modified chemistry largely depends on the Tm needed to melt this structure. See Chapter 5 for more details.

Type	Length	Count	Positions	Sequence	
Direct	**34**	**2**	**112,220**	**GCGCGCACTC CGAAATTGTG TTGACACAGT CTCC**	
Direct	29	2	151,934	GCGGCACCCT GGTCACTGTC TCCTCTGAT	
Direct	11	2	322,694	ACATGCACTG G	
Direct	10	2	466,648	CTGAAGATTT	
Invert	**47**	**2**	**199,200**	**GTGTCAACAC AATTTCGGAG TGCGCGCACT CCGAAATTGT GTTGACA**	
Invert	27	2	112,199	GCGCGCACTC CGAAATTGTG TTGACAC	
Invert	12	2	190,947	GAGGAGACAG TG	
Invert	12	2	164,190	CACTGTCTCC TC	
Invert	11	2	394,395	CTGGAATTCC A	
Palind	**48**	**2**	**199,246**	**GTGTCAACAC AATTTCGGAG TGCGCGCACT CCGAAATTGT GTTGACAC**	
Palind	46	2	200,245	TGTCAACACA ATTTCGGAGT GCGCGCACTC CGAAATTGTG TTGACA	
Palind	44	2	201,244	GTCAACACAA TTTCGGAGTG CGCGCACTCC GAAATTGTGT TGAC	
Palind	42	2	202,243	TCAACACAAT TTCGGAGTGC GCGCACTCCG AAATTGTGTT GA	
Palind	40	2	203,242	CAACACAATT TCGGAGTGCG CGCACTCCGA AATTGTGTTG	
Palind	38	2	204,241	AACACAATTT CGGAGTGCGC GCACTCCGAA ATTGTGTT	
Palind	36	2	205,240	ACACAATTTC GGAGTGCGCG CACTCCGAAA TTGTGT	
Palind	34	2	206,239	CACAATTTCG GAGTGCGCGC ACTCCGAAAT TGTG	
Palind	32	2	207,238	ACAATTTCGG AGTGCGCGCA CTCCGAAATT GT	
Palind	30	2	208,237	CAATTTCGGA GTGCGCGCAC TCCGAAATTG	
Palind	28	2	209,236	AATTTCGGAG TGCGCGCACT CCGAAATT	
Palind	26	2	210,235	ATTTCGGAGT GCGCGCACTC CGAAAT	
Palind	24	2	211,234	TTTCGGAGTG CGCGCACTCC GAAA	
Palind	22	2	212,233	TTCGGAGTGC GCGCACTCCG AA	
Palind	20	2	213,232	TCGGAGTGCG CGCACTCCGA	
Palind	18	2	214,231	CGGAGTGCGC GCACTCCG	
Palind	12	2	394,405	CTGGAATTCC AG	
Palind	10	2	345,354	CCTGGCCAGG	
Palind	10	2	894,903	TACTATAGTA	
Palind	10	2	395,404	TGGAATTCCA	

Figure 8-10. The potentially difficult to sequence motifs detected by Examine Repeats module. The presence of long inverted repeat structure requires using modified chemistry to obtain clean sequencing trace in this region. On the other hand the direct repeat and palindromes should not present any problems as long as the primers are not located in the region of a repeat or palindromes.

and then checked against each remaining position. The segment is then incremented one base and the process repeated. The primary difference between the inverted and direct repeats is comparison of the segment itself versus the reverse complement. Palindromes are detected by checking the segment against its own reverse complement.

Homopolymers, dinucleotide repeats, and trinucleotide repeats are found by examining the template for any 1, 2, or 3 ordered base patterns along the entire length. Dinucleotide non-repeats are detected by finding stretches of two bases that are either not dinucleotide repeats or composed of at least 25% of each of the two bases. When the same repeat stretch is detected multiple times by the sliding window, these are then collapsed.

The direct, inverted, and palindrome repeat detection code is executed in parallel using Perl's fork feature and the results are reassembled for return to LIMS in an XML document. Given the CPU resources needed to return the sliding window match results in real time, parallel execution is a necessity.

The size of repeat segments is highly configurable. The size of the segment used for direct, inverted, and palindromic repeats can be set; the threshold for the remaining repeat types is also a parameter.

Summary

To efficiently deal with various types of sequencing requests encountered in any typical DNA sequencing facility (at Level 2 and especially at Level 3), all components of the process flow need to be optimized. Experienced personnel, robust protocols, instrumentation, and bioinformatics tools need to be integrated and coordinated for maximum efficiency. Although the volume of requests containing difficult templates is on the order of 10% (in our case), the amount of effort needed to deliver good quality data in a reasonable time period is disproportionately higher compared to a standard sequencing request. By our observation, the amount of effort (i.e., time and reagents) needed to sequence some difficult templates could be three to five times higher compared to a "normal" template if the core facility staff does not have advanced protocols and bioinformatics tools. Incorporating GC and Examine Repeats modules allows us (at the Wyeth core DNA facility) to predict in advance many different structures that most of the time have deleterious effects on the sequence quality. For GC module content, we have coupled a level of GC content with experimentally determined chemistry, and our LIMS system automatically suggests it while creating the sample sheet (ABI sequencer input file). Currently, we are working to incorporate appropriate chemistry with other types of difficult templates into our LIMS.

Through the development of advanced sequencing protocols (see Chapters 1 and 2) and application of sophisticated bioinformatics tools we are able to significantly reduce the time and effort while dealing with many types of difficult regions. However, much broader data set will be necessary to fully elucidate the correlation between the type of difficult region and a successful chemistry needed to read through such a region. Although our LIMS system is capable to detect many most common types of difficult regions that impede clean read-through, much more detailed bioinformatics tools may be needed to detect more nuanced sequence patterns, especially those that immediately precede and follow a difficult region.

References

1. *ABI PRISM® BigDye™ Terminator v3.1 Cycle Sequencing Kit.* Protocol. Part number 4337035 Rev. A. Foster City, CA: Applied Biosystems; 2002.
2. Adams, P.S., Dolejsi, M.K., Hardin, S., et al. 1996. DNA sequencing of a moderately difficult template: evaluation of the results from a Thermus thermophilus unknown test sample. *BioTechniques* 21: 678.
3. Adams, P.S., Dolejsi, M.K., Hardin, S., et al. 1997. Effects of DMSO, thermocycling and editing on a template with a 72% GC rich area: results from the 2nd Annual ABRF Sequencing Survey demonstrate that editing is the major factor for improving sequencing accuracy. Ninth International Genome Sequencing and Analysis Conference. *Microb Comp Genomics* 2: 198 (abstract).
4. Adams, P.S., Dolejsi, M.K., Grills, G., et al. 1999. An analysis of techniques used to improve the accuracy of automated DNA sequencing of a GC-rich template: results from the 2nd ABRF DNA Sequence Research Group Study. Available at: http://www.abrf.org; search: 2nd ABRF DNA Sequence Research Group Study. Accessed November 2007.
5. Brummelkamp, T.R., Bernards, R., and Agami, R. A system for stable expression of short interfering RNAs in mammalian cells. *Science* 2002; 296: 550–553.
6. Burgett, S.G., and Rosteck, P.R. Jr. Use of dimethyl sulfoxide to improve fluorescent, Taq cycle sequencing. In Adams, M.D., Fields, C., and Venter, J.C., eds. *Automated DNA Sequencing and Analysis.* New York: Academic Press; 1994: 211–215.
7. Elbashir, S.M., Harborth, J., Weber, K., and Tuschl, T. Analysis of gene function in somatic mammalian cells using small interfering RNAs. *Methods* 2002; 26: 199–213.
8. Fielding, R.T. *Architectural Styles and the Design of Network-Based Software Architectures.* Doctoral dissertation. Irvine, CA: University of California, Irvine; 2000.
9. Hammond, S.M., Caudy, A.A., and Hannon, G.J. Posttranslational gene silencing by double-stranded RNA. *Nat Rev Genet* 2001; 2:110–119.

10. International Human Genome Sequencing Consortium. Initial sequencing and analysis of the human genome. *Nature* 2001; 409: 860–921.
11. Kieleczawa, J. Sequencing of difficult DNA templates. In: Kieleczawa, J. (ed). *DNA Sequencing: Optimizing the Process and Analysis.* Sudbury, MA: Jones and Bartlett Publishers; 2005: 27–34.
12. Kieleczawa, J. Simple modifications of the standard DNA sequencing protocol allow for sequencing through siRNA hairpins and other repeats. *J Biomol Tech* 2005; 16: 220–223.
13. Kieleczawa, J. Controlled heat-denaturation of DNA plasmids. In: Kieleczawa, J. (ed). *DNA Sequencing: Optimizing the Process and Analysis.* Sudbury, MA: Jones and Bartlett Publishers; 2005: 1–10.
14. Kieleczawa, J., and Bajson, K. Evaluation of methods for cleanup of DNA sequencing reactions. In: Kieleczawa, J. (ed). *DNA Sequencing II: Optimizing the Preparation and Clean Up.* Sudbury, MA: Jones and Bartlett Publishers; 2006: 219–240.
15. Kieleczawa, J., and Wu, P. Prolonged storage of plasmid DNAs under different conditions: effects on plasmid integrity, spectral characteristics, and DNA sequence quality. In: Kieleczawa, J. (ed). *DNA Sequencing II: Optimizing the Preparation and Clean Up.* Sudbury, MA: Jones and Bartlett Publishers; 2006: 259–274.
16. Kieleczawa, J., Li, T., and Wu, P. Preparation of difficult DNA templates using seven different commercial methods. In: Kieleczawa, J. (ed). *DNA Sequencing II: Optimizing the Preparation and Clean Up.* Sudbury, MA: Jones and Bartlett Publishers; 2006: 1–14.
17. Kieleczawa, J. Fundamentals of sequencing of difficult templates—an overview. *J Biomol Tech* 2006; 17: 207–217.
18. Koffman, D.K., and Sookdeo, H. DNA sequencing database: a flexible LIMS for DNA sequencing analysis. In: Kieleczawa, J. (ed). *DNA Sequencing: Optimizing the Process and Analysis.* Sudbury, MA: Jones and Bartlett Publishers; 2005: 143–156.
19. Sanger, F., Nicklen, S., and Coulson, A.R. DNA sequencing with chain-terminating inhibitors. *Proc Natl Acad Sci U S A* 1977; 74: 5463–5467.
20. *Sequence Finishing Kit.* Product Code 25-6401-01. Piscataway, NJ: GE Healthcare; 2003.
21. Venter, J.C., Adams, M.D., Myers, E.W., Li, P.W., Mural, R.J., Sutton, G.G., Smith, H.O., Yandell, M., et al. The Sequence of the human genome. *Science* 2001; 291: 1304–1351.

Index